AF680586

The Residential HVAC Masterclass

The Complete Guide to Understanding, Diagnosing, and Maintaining Modern Heating and Cooling Systems

By Jim Cox

DISCLAIMER & SAFETY SCOPE

CRITICAL NOTICE TO THE READER: PLEASE READ BEFORE PROCEEDING

This publication is designed to provide accurate and authoritative educational information regarding the subject matter covered—specifically, residential climate systems, preventative maintenance, and basic low-voltage and high-voltage electrical diagnostics.

However, HVAC equipment utilizes dangerous electrical voltages (120\text{ VAC} to 240\text{ VAC}), combustible gas pressures, hazardous chemical refrigerants, and high-speed moving machinery. Handling these components improperly carries an inherent risk of severe electrical shock, fatal electrocution, explosion, fire, property damage, and personal injury.

THE STRUCTURAL BOUNDARIES OF THIS HANDBOOK

To ensure absolute safety, legal compliance, and accessibility for beginners, this book enforces a strict technical boundary. This handbook is strictly diagnostic and preventative in nature. ` ` `

WHAT THIS BOOK TEACHES	WHAT THIS BOOK EXCLUDES
(Safe, Legal DIY Actions)	(EPA-Regulated & High-Risk)
✓ Safe Multimeter Operation	✗ Refrigerant Charging & Pressures
✓ 24V Control Circuit Diagnostics	✗ Refrigerant Recovery & Reclamation
✓ Safety Limit Switch Continuity Tests	✗ Compressor Replacement & Wiring
✓ Dual-Run Capacitor Bench Testing	✗ Torch Brazing Sealed Copper Lines
✓ Contact Resistance & Voltage Drops	✗ High-Pressure Leak Detection
✓ Deep Chemical Condenser Cleaning	✗ Any EPA Section 608 Regulated Work
NO SPECIALIZED EPA LICENSING NEEDED	REQUIRES EPA LICENSE & TECH TOOLS

LEGAL LIMITATION OF LIABILITY

The structural procedures, diagnostic tables, and mechanical step-by-step frameworks contained within this manuscript have been thoroughly researched and verified for technical accuracy.

Nevertheless, neither the author nor the publisher assumes any responsibility or legal liability for any damages, injuries, financial losses, or equipment failures incurred as a direct or indirect result of following the information in this book.

By continuing past this page, you explicitly acknowledge and agree to the following conditions:

* **Local Codes Take Precedence:** You must always cross-reference any diagnostic step with your specific local electrical codes, mechanical codes, and original manufacturer instructions.

* **The Right to Stop:** If at any point during a diagnostic sequence your meter readings do not match the expected parameters outlined in this book, or if you feel uncomfortable working around a live component, **you must immediately stop, shut down the system breaker, and consult a licensed HVAC professional.**

* **The Safety Mandate:** You agree to always wear the mandatory Personal Protective Equipment (PPE)—including ANSI-approved safety glasses and insulated, rubber-coated mechanics gloves—whenever performing any physical test detailed in this guide.

> **A Note on EPA Section 608 Federal Law:** Under Title VI of the Clean Air Act, it is a federal violation to open, cut, adjust, or connect gauges to the sealed refrigerant circuit of an air conditioner or heat pump without holding an active EPA Section 608 Certification. Violations carry severe civil penalties. This book will **never** ask you to breach the sealed refrigeration circuit or handle refrigerant in any capacity. Everything taught herein is 100% legal, safe, and accessible for an un-certified __homeowner working on their own personal property.__

TABLE OF CONTENTS

CHAPTER 1 — The HVAC Mindset: Thinking Like a Technician

Why HVAC Isn't About Guessing — It's About Following the Sequence

Most homeowners think HVAC work is mysterious.
Technicians know the truth: **HVAC systems are predictable.**
They follow the same steps, in the same order, every single time.

If you understand the order — the *sequence of operation* — you can diagnose almost anything without guessing, without replacing random parts, and without wasting money.

This chapter teaches you how to think like a technician, not a parts-changer.

1.1 HVAC Is a Logic Puzzle, Not a Mystery

Every furnace, air conditioner, and heat pump is built around the same idea:

A thermostat makes a request.
The system responds in a specific order.
If the order breaks, the system stops.

That's it.
HVAC is a chain reaction.

If you know the chain, you know where it broke.

1.2 The Three Systems Inside Every HVAC Unit

Every residential HVAC system is really three systems working together:

1. The Electrical System

This includes:

- 24-volt transformer
- Thermostat wiring
- Control board
- Relays
- Safety switches
- Capacitors
- Motors

This is the "brain" of the system.

2. The Airflow System

This includes:

- Blower motor
- Evaporator coil
- Return ducts
- Supply ducts
- Filters
- Outdoor condenser fan

This is the "lungs" of the system.

3. The Fuel or Refrigeration System

Depending on the equipment:

- Gas valve
- Burners
- Igniter
- Flame sensor
- Compressor
- Refrigerant circuit (diagnostics only — no EPA-restricted work)

This is the "muscle" of the system.

When something fails, it's always one of these three.

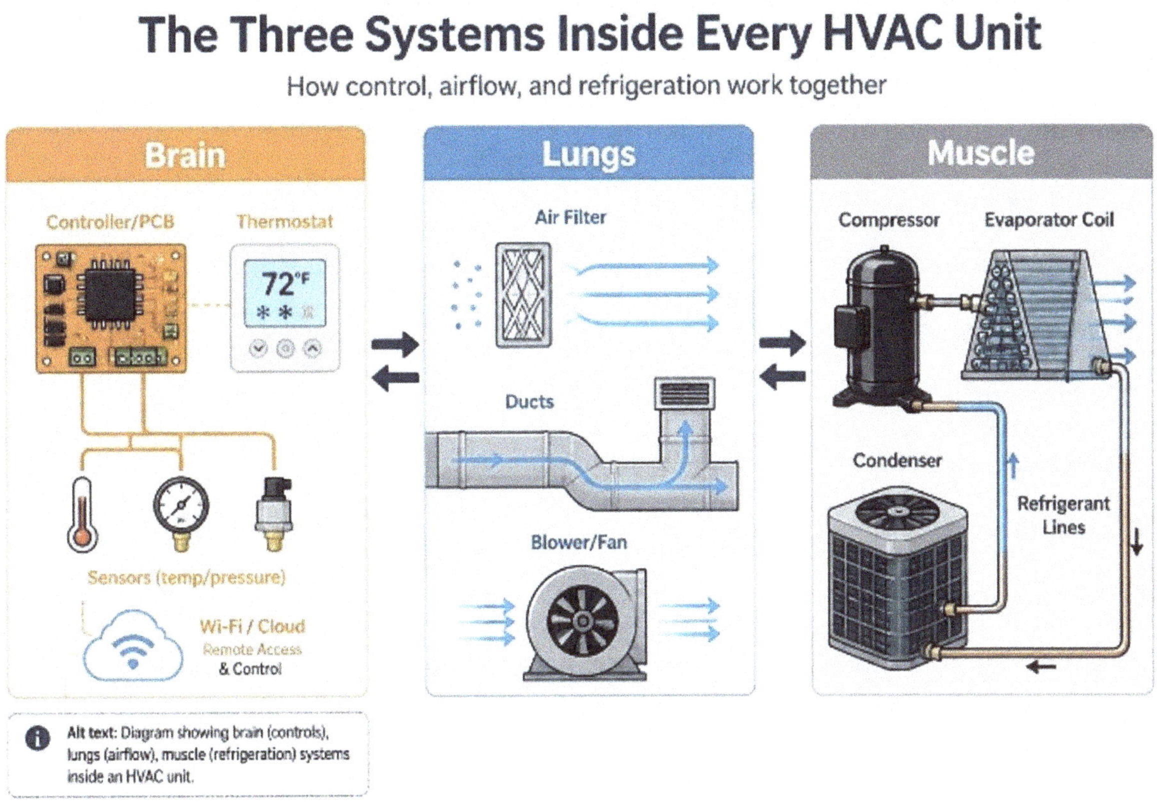

1.3 Why Homeowners Struggle With HVAC

Most people try to diagnose HVAC systems by symptoms:

- "It's blowing warm air."
- "The furnace won't start."
- "The AC hums but doesn't run."

Symptoms are helpful, but they don't tell you *why*.

Technicians don't diagnose symptoms.
They diagnose **where the sequence stopped**.

1.4 The Technician's Advantage: Sequence of Operation

Every HVAC system follows a predictable order.

For example, in heating mode:

1. Thermostat calls for heat
2. Inducer motor starts
3. Pressure switch closes
4. Igniter glows
5. Gas valve opens
6. Flame sensor verifies flame
7. Blower motor starts

If the furnace fails at Step 3, you know the pressure switch or inducer is the issue.
If it fails at Step 5, you know the gas valve isn't opening.

This is why technicians seem like magicians — they're not guessing.
They're following the chain.

1.5 Why This Book Works

This book is built around the same method technicians use:

- Understand the sequence
- Test the electrical path
- Verify airflow
- Check safety switches
- Confirm the system's logic

You'll learn:

- How to test 24V and 120V safely
- How to test capacitors
- How to read control board flash codes
- How to diagnose airflow problems
- How to clean and maintain equipment
- How to wire thermostats correctly

And you'll do it **without**:

- Handling refrigerant
- Opening sealed systems
- Performing EPA-restricted tasks

This keeps everything safe, legal, and accessible.

1.6 The Three Questions Every Technician Asks

Whenever a system fails, a technician asks:

1. Do we have power?

24V?
120V?
240V?

2. Do we have airflow?

Filter?
Blower?
Coils?
Ducts?

3. Do we have a complete sequence?

Where did the chain stop?

If you can answer these three questions, you can diagnose 90% of residential HVAC problems.

1.7 The Technician's Mindset

To think like a technician, you must:

Stay calm

HVAC failures often happen in extreme weather.
Don't rush.

Follow the sequence

Never skip steps.
Never jump ahead.

Test, don't guess

Meters don't lie.
Symptoms do.

Respect safety

HVAC involves:

- High voltage
- Gas
- Moving parts
- Pressure

You'll learn how to work safely in every chapter.

1.8 What You'll Learn Next

In the next chapters, you'll learn:

- The full heating sequence
- The full cooling sequence
- How the thermostat communicates
- How the 24V control system works

Once you understand these, you'll be able to diagnose:

- No-heat
- No-cool
- Short cycling
- Blower failures
- Ignition failures
- Capacitor failures
- Contactor failures
- Thermostat issues
- Airflow restrictions

This is where the book starts to feel like having a technician standing next to you.

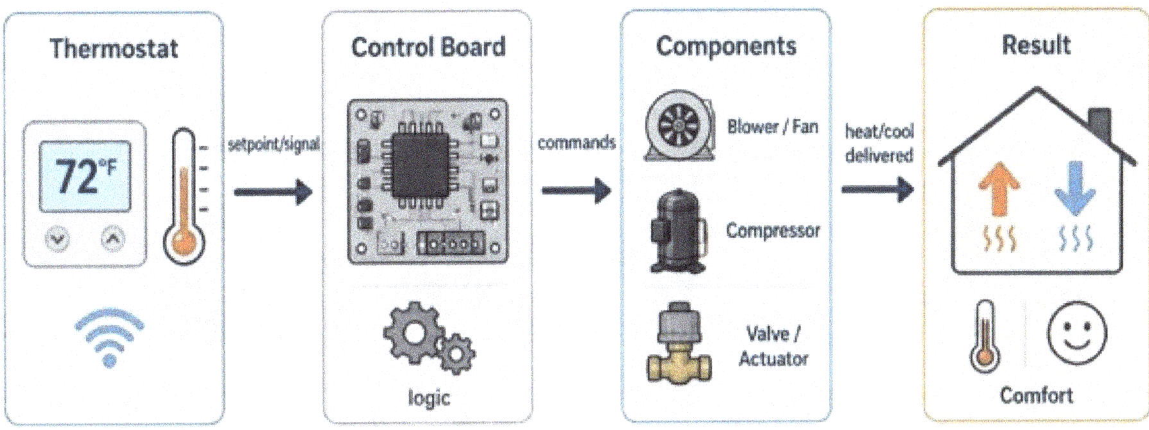

Alt text: Flowchart showing thermostat to control board to components to result.

CHAPTER 2 — The Sequence of Operation (Heat Mode)

The Most Important Concept in All of HVAC Diagnostics

If you learn only one thing from this entire book, let it be this chapter.

The sequence of operation is the technician's roadmap. It tells you exactly what the furnace should do, in the exact order it should do it. When something goes wrong, you simply find the step where the sequence stopped—and that's your problem.

No guessing. No replacing random parts. No frustration.

Let's walk through the entire heating sequence like a pro.

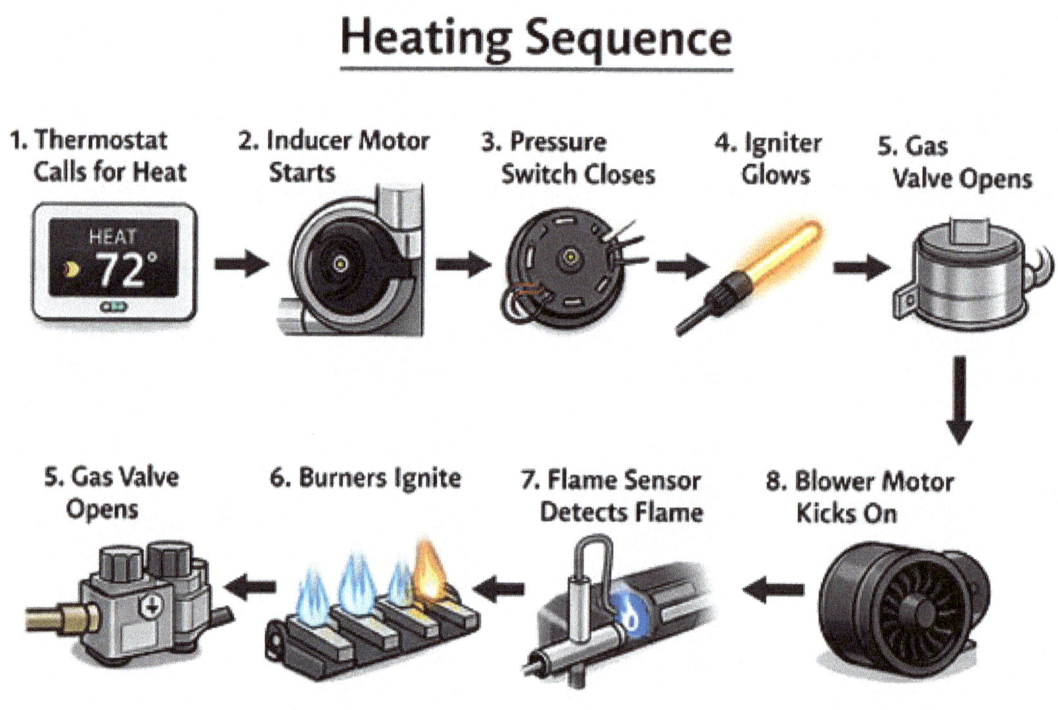

2.1 What "Sequence of Operation" Really Means

Every furnace—regardless of brand, size, or efficiency level—follows the same basic chronological steps when the thermostat calls for heat.

Think of it like a row of dominoes. If one domino doesn't fall, the rest of the steps never happen. Your job as a diagnostician isn't to fix the whole furnace; it's simply to find the exact domino that failed to tip over.

2.2 Step-By-Step: What Happens When You Turn Up the Thermostat

Below is the standard, modern 9-step heating sequence used by virtually every gas furnace on the market today. We'll break down each step, explain what should happen, and show you exactly what it means when the chain breaks.

STEP 1 — Thermostat Calls for Heat (W \rightarrow R Closes)

When you raise the temperature setting above the current room temperature:

- The thermostat closes an internal switch, connecting the W (heating call) circuit to R (24V hot power).
- This sends a continuous 24-volt AC signal down to the furnace control board.

If this step fails: You will find no 24V power reading between terminals W and C at the furnace control board. Common culprits include a dead thermostat battery, a broken thermostat wire, a blown low-voltage 3-amp or 5-amp fuse on the board, or a failed 24V transformer.

STEP 2 — Safety Circuit Verification (The Gatekeeper Step)

Before the control board passes power to any motors, it performs a lightning-fast self-test of the safety limit loop. It checks to make sure the following switches are closed:

- High-Limit Switch (monitors heat exchanger temperature)
- Rollout Switches (monitors for stray flames escaping the burners)
- Auxiliary Limit Switches (monitors internal blower cabinet temperatures)

If this step fails: If any of these safety switches are stuck open from a previous malfunction or overheating event, the sequence halts instantly. On many modern boards, the indoor blower motor will turn on and run continuously to cool the system down, but ignition will be locked out.

STEP 3 — Inducer Motor Starts

Once safeties are verified closed, the control board closes a relay to send 120V household power to the combustion inducer draft motor. This small, dedicated fan:

- Pre-purges the heat exchanger of any leftover gases.
- Pulls fresh combustion air into the burners.
- Creates a negative static pressure to draw dangerous exhaust safely out through the vent pipe or chimney.

If this step fails: The small exhaust fan won't spin. Look for a seized inducer motor bearing, a dead motor winding, a blockage inside the flue pipe, or a failed relay on the control board failing to send 120V to the motor.

STEP 4 — Pressure Switch Closes

As the inducer motor reaches full operating speed, the suction it creates is measured by a small, flexible diaphragm inside the pressure switch. This vacuum pulls the switch closed, sending a signal back to the board that says: *"The vent path is clear. It is safe to ignite."*

If this step fails: This is the #1 cause of a furnace that "tries to start but never ignites." The inducer will run indefinitely, but nothing else happens. Look for a clogged collection port where the vinyl tube meets the plastic inducer housing, a cracked or rotting pressure tube, a blocked vent pipe, or a failed pressure switch diaphragm.

STEP 5 — Hot Surface Igniter Glows

Once the pressure switch closes and proves safe draft, the board sends power to the Hot Surface Igniter (HSI). The igniter will begin to glow a bright, intense orange—usually for a window of 10 to 20 seconds to reach ignition temperature.

- Tech Note on Voltage: While classic silicon carbide igniters run on standard 120 VAC, many modern high-efficiency units utilize rugged silicon nitride igniters designed to run on lower voltages, such as 24 VAC or 80 VAC regulated directly by the control board. Always check your unit's wiring schematic before testing.

If this step fails: The igniter remains completely dark. Igniters are incredibly fragile and prone to cracking over time due to thermal stress. Look for a physically cracked or burned igniter element, a loose wiring plug, or a failure on the control board to send the correct voltage.

STEP 6 — Gas Valve Opens

When the igniter reaches its peak temperature, the control board sends a 24V signal to the gas valve solenoids. The valve clicks open, allowing gas to flow into the manifold and across the burner assembly. You should distinctly hear a soft mechanical *click* followed immediately by the rushing sound of gas escaping.

If this step fails: The igniter glows brilliantly, but goes out without a flame ever appearing. Look for an closed manual gas shut-off valve outside the furnace cabinet, no 24V signal arriving at the valve terminals, or a mechanically seized gas valve.

Alt text: Cutaway of gas furnace burner rack showing burners, crossover channels, hot surface igniter, flame sensor, gas manifold.

Cutaway: Gas furnace burner rack components

STEP 7 — Burners Ignite

The gas flowing out of the burners travels across the glowing hot surface igniter, resulting in a smooth, quiet "whoosh" as the entire burner assembly catches fire.

If this step fails: A delayed ignition occurs. If the gas doesn't light instantly and instead pools for a few seconds before popping loudly, shut the system down immediately. Common causes include dirty burner ports, misaligned burner crossover brackets, or low manifold gas pressure.

STEP 8 — Flame Sensor Verifies Flame

This is a critical safety step. Within 1 to 3 seconds of the gas valve opening, the furnace must prove that fire actually exists. The control board pushes a small AC voltage out to the flame sensor rod. When engulfed in fire, the physics of the flame converts this into a microscopic DC current (measured in microamps, \mu A) that flows back to the furnace ground.

If this step fails: The burners will light perfectly, but shut completely off after exactly 3 to 5 seconds. If this happens 3 consecutive times, the board locks out. Cleaning a microscopic layer of clear silica oxidation off the flame sensor with a piece of light abrasive pad or steel wool is one of the easiest, cheapest, and most common DIY HVAC fixes on the planet.

STEP 9 — Main Indoor Blower Motor Starts

The burners are lit and generating intense heat inside the heat exchanger chambers. To prevent the steel from overheating, the control board waits a predetermined time (usually 30 to 45 seconds) and then engages the main indoor blower motor to push cozy, warm air out through your home's ductwork.

If this step fails: The burners will run for a few minutes, but the house won't get warm. Eventually, the air inside the cabinet gets so hot that it trips the High-Limit Switch (Step 2), cutting off the gas valve. Look for a bad blower motor, a failed run capacitor, or a broken blower relay on the circuit board.

2.3 What Happens When the Thermostat Reaches Setpoint?

Once your living space reaches the desired temperature setting on the wall:

1. The thermostat opens the W \right arrow R connection, removing the 24V signal from the board.
2. The control board immediately cuts power to the gas valve, causing the burners to extinguish.
3. The inducer motor runs for a brief 5 to 15-second "post-purge" to clear residual combustion fumes.
4. The main indoor blower motor continues to run for an additional 60 to 120 seconds. This "heat off-delay" is vital—it harvests the remaining heat out of the heat exchanger, maximizing efficiency and protecting the steel from cracking due to sudden heat stagnation.

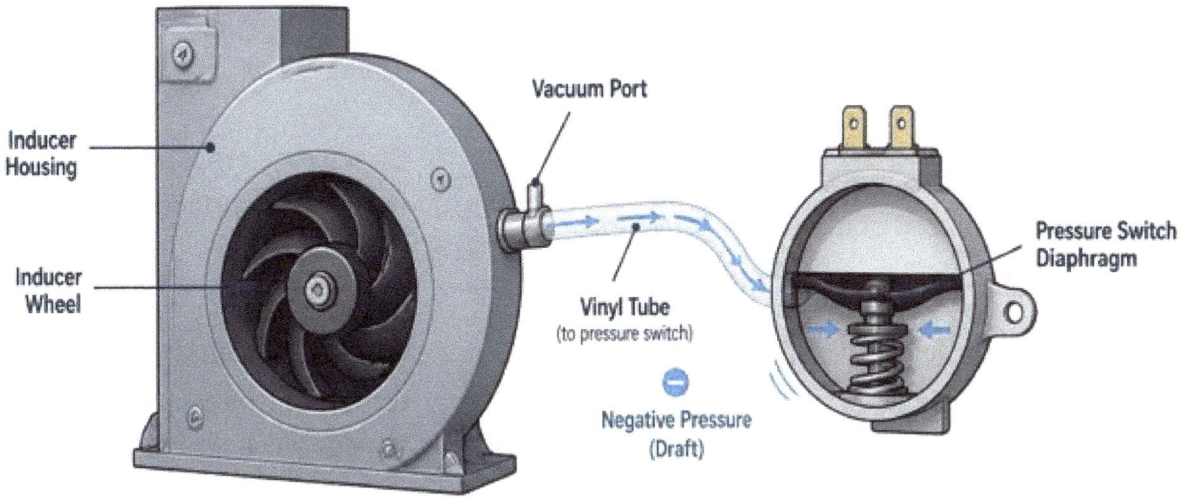

Pressure switch senses negative pressure (draft) from inducer.

Alt text: Side view of inducer housing with vacuum port connected via vinyl tube to a pressure switch; arrows show negative pressure pulling the switch diaphragm inward.

→ Negative Pressure (suction)

⊖ Negative Pressure (−)

2.4 Diagnostic Blueprint: Where Did the Chain Break?

Because every furnace failure is simply a broken step in the sequence, you can map symptoms directly to the faulty component without guessing.

If you notice this symptom...	It means the sequence stopped at...	Focus your diagnostics here...
The thermostat clicks, but the furnace cabinet remains completely silent.	Step 1 or 2: Thermostat Call or Safety Limits	Check for a blown 3A board fuse, open safety switches, or a dead 24V transformer.
The small exhaust fan runs continuously, but the igniter never glows.	Step 4: Pressure Switch Verification	Check for a clogged pressure port tube, blocked exhaust pipe, or a failed switch.

If you notice this symptom...	It means the sequence stopped at...	Focus your diagnostics here...
The exhaust fan runs, but the igniter stays dark and cold.	Step 5: Igniter Activation	Check for a cracked or burned hot surface igniter element.
The igniter glows bright orange, but no fire ever appears.	Step 6 or 7: Gas Flow & Ignition	Verify the gas line valve is turned on and test for 24V at the gas valve terminals.
The burners light up beautifully, but snap completely off after 3 to 5 seconds.	Step 8: Flame Proving	Clean the flame sensor rod of invisible silica buildup.
The furnace heats up, but shuts down after a few minutes while blowing lukewarm air.	Step 9: Airflow Delivery	Check for a clogged air filter, blocked return vents, or a bad blower motor capacitor.

2.5 What You'll Learn Next

Now that you have mastered the logical roadmap of how a heating system thinks, we need to look at the other side of the coin. In Chapter 3, we dive into the Sequence of Operation (Cooling Mode).

Cooling mode relies on the exact same structural logic, but introduces a new cast of characters: contactors, dual-run capacitors, compressors, and condenser fans. Let's turn the page.

CHAPTER 3 — The Sequence of Operation (Cooling Mode)

How an Air Conditioner Thinks — And Why It Fails in Predictable Ways

If Chapter 2 taught you how a furnace thinks, this chapter teaches you how an **air conditioner** thinks.
Cooling mode is simpler than heating mode — fewer safety switches, fewer moving parts — but the logic is just as strict.

When the thermostat calls for cooling, the system follows a **fixed electrical path**.
If any part of that path breaks, the AC stops working.

Once you understand this sequence, diagnosing AC problems becomes shockingly easy.

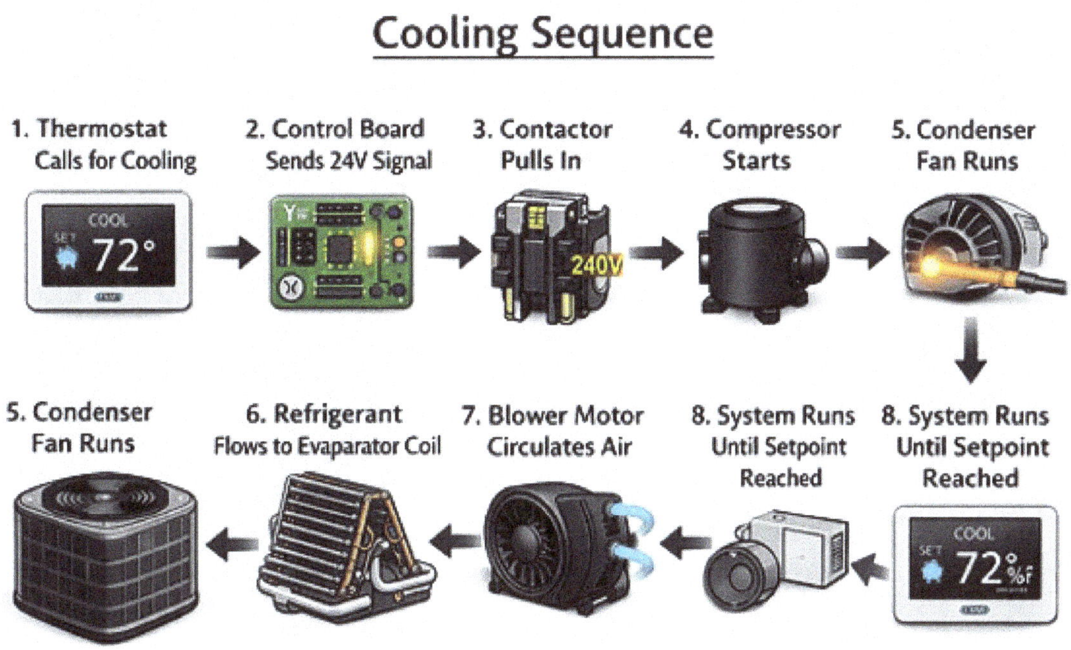

3.1 The Cooling System Is Really Two Systems Working Together

When you turn on the AC, you're activating:

1. The Indoor System

- Blower motor
- Evaporator coil
- Control board
- 24V transformer
- Thermostat wiring

2. The Outdoor System

- Compressor
- Condenser fan motor
- Dual-run capacitor
- Contactor
- High-voltage disconnect

These two systems communicate through a **single 24-volt signal**: the **Y** wire.

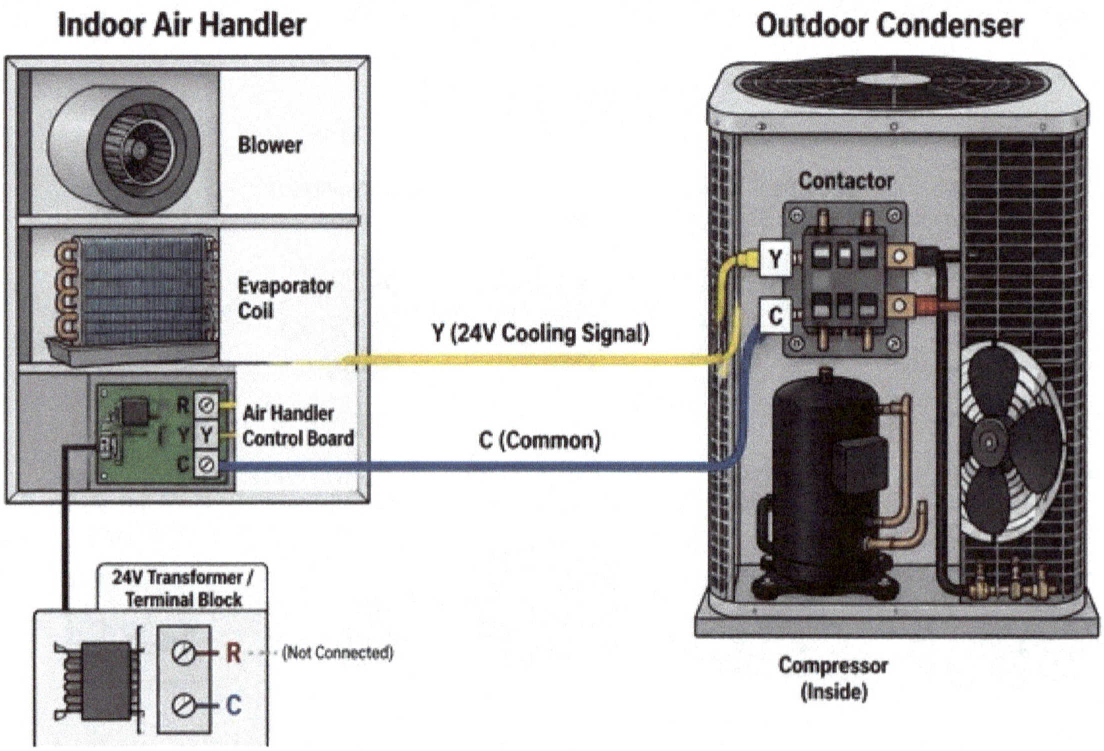

3.2 The Cooling Sequence of Operation (Step-By-Step)

Here is the complete cooling sequence used by nearly every residential split system.

STEP 1 — Thermostat Calls for Cooling (Y → R Closes)

When you lower the thermostat temperature:

- The thermostat closes **Y** to **R**
- This sends 24V to the furnace control board
- The board sends 24V outside to the **contactor coil**

If this step fails:

- No 24V between R and Y
- Bad thermostat
- Broken thermostat wire
- Blown low-voltage fuse
- Failed transformer

This is the #1 cause of "AC won't turn on."

STEP 2 — Indoor Blower Motor Starts

The control board energizes the **G** circuit, starting the indoor blower.

This pushes warm indoor air across the evaporator coil.

If blower doesn't start:

- Bad blower motor
- Bad capacitor (PSC motors)
- Failed ECM module (ECM motors)
- Bad relay on control board
- Clogged evaporator coil causing limit trips

A non-running blower can freeze the evaporator coil solid.

STEP 3 — Outdoor Contactor Pulls In

The 24V signal energizes the **contactor coil**.

This pulls in the contactor, connecting **240V power** to:

- The compressor
- The condenser fan motor

If the contactor doesn't pull in:

- No 24V at contactor
- Bad contactor coil
- Broken low-voltage wire
- Safety switch open (rare)

This is the #1 diagnostic point for AC technicians.

STEP 4 — Compressor Starts

Once the contactor closes, the compressor receives 240V.

It begins pumping refrigerant through the system.

If compressor hums but won't start:

- Bad capacitor (most common)
- Weak start winding
- Failed compressor
- Low voltage at contactor

A humming compressor is almost always a **bad dual-run capacitor**.

STEP 5 — Condenser Fan Motor Starts

The outdoor fan pulls air through the condenser coil, removing heat from the refrigerant.

If fan doesn't start:

- Bad capacitor
- Bad fan motor
- Seized bearings
- Failed contactor

If the fan doesn't run, the compressor will overheat and shut down.

STEP 6 — Indoor Evaporator Coil Absorbs Heat

Warm indoor air passes over the evaporator coil.

The coil absorbs heat and moisture, cooling and dehumidifying the air.

If coil freezes:

- Dirty filter
- Dirty evaporator coil
- Low airflow
- Blower not running
- Low refrigerant (diagnosis only — no EPA work)

Airflow problems cause 80% of freeze-ups.

STEP 7 — Condenser Coil Rejects Heat

The outdoor coil releases heat into the outside air.

If condenser coil is dirty:

- AC runs constantly
- Poor cooling
- High energy bills
- Compressor overheating

A dirty condenser coil is the #1 cause of poor cooling performance.

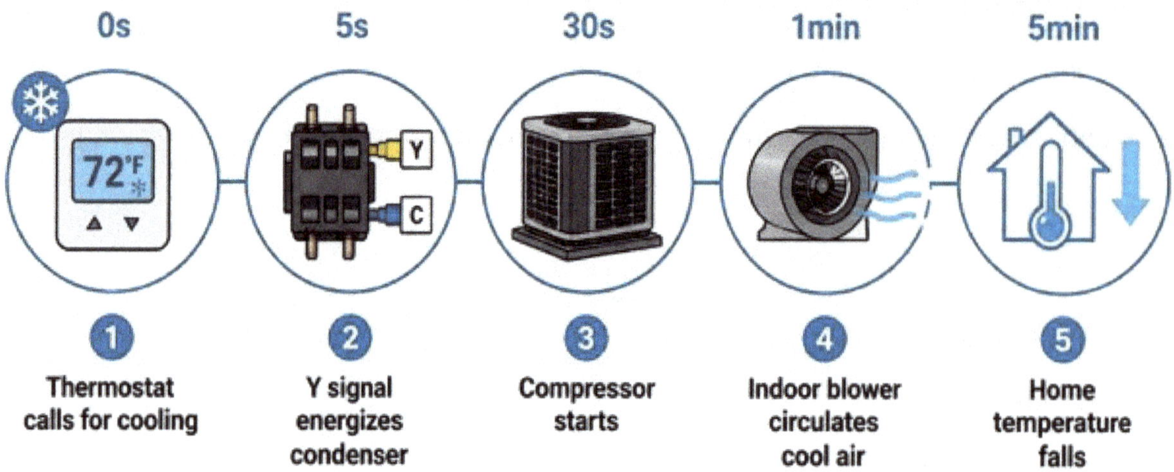

STEP 8 — System Runs Until Setpoint Is Reached

When the home reaches the target temperature:

- Thermostat opens Y → R
- Contactor drops out
- Compressor stops
- Outdoor fan stops
- Indoor blower runs for 60–90 seconds
- System returns to standby

This "after-blow" removes remaining cold air from the coil.

3.3 The Cooling Sequence Diagnostic Blueprint

Because cooling mode operates on a strict electrical chain reaction, you can look at exactly where the system stops running and map it directly to a failed component or signal.

Use this master flowchart matrix to track the sequence like a pro:

Sequence Stage	Circuit & Voltage	What Is Happening Mechanically	If the System Stops Here, Troubleshoot...
1. Thermostat Call	Low Voltage (24 VAC) Y \rightarrow R Closes	The thermostat bridges 24V power to the **Y** terminal, sending a cooling request to the indoor furnace/air handler control board.	Dead thermostat batteries, a broken thermostat wire, a blown low-voltage board fuse, or a failed 24V transformer.
2. Control Board Relay	Low Voltage (24 VAC) Y and C to Outdoor Unit	The indoor board registers the call, immediately engages the indoor blower relay, and routes 24V power out to the outdoor condenser cabinet.	A failed indoor blower relay or a loose/corroded wiring terminal connection at the control board.
3. Contactor Coil	Low Voltage (24 VAC) At Outdoor Unit	The incoming 24V signal hits the magnetic coil of the outdoor contactor switch, creating an electromagnetic field.	A burned-out contactor coil or a broken low-voltage wire broken by a lawnmower or weed-whacker outside.
4. Contactor Closes	High Voltage (240 VAC)	The magnetic field pulls the heavy metal contacts down, allowing 240V house power to	Pitted, burned, or bug-infested contactor points (a massive

Sequence Stage	Circuit & Voltage	What Is Happening Mechanically	If the System Stops Here, Troubleshoot...
	Line to Load Side	cross from the incoming line side to the load side.	cause of summer AC failures).
5. Compressor Start	High Voltage (240 VAC) Hermetic Circuit	High voltage rushes into the compressor. The compressor begins pumping and compressing vapor refrigerant through the sealed system.	A blown dual-run capacitor (HERM terminal), low line voltage, or a mechanically seized compressor motor.
6. Outdoor Fan Start	High Voltage (240 VAC) Fan Circuit	Simultaneously, 240V rushes into the condenser fan motor, spinning the blades to pull outdoor air across the aluminum fins.	A blown dual-run capacitor (FAN terminal), seized motor bearings, or a burned-out fan motor winding.
7. Indoor Heat Absorption	Thermal Transfer Evaporator Coil	The indoor blower pushes warm house air across the freezing cold indoor evaporator coil, absorbing heat and sweating out humidity.	A frozen evaporator coil, a severely clogged air filter, a broken indoor blower motor, or a low refrigerant charge.

Sequence Stage	Circuit & Voltage	What Is Happening Mechanically	If the System Stops Here, Troubleshoot...
8. Outdoor Heat Rejection	Thermal Transfer Condenser Coil	The boiling hot vapor refrigerant passes through the outdoor coil, where the fan rejects the captured indoor heat into the outside air.	A heavily impacted or dirty outdoor condenser coil coated in grass clippings, dirt, or cottonwood fuzz.

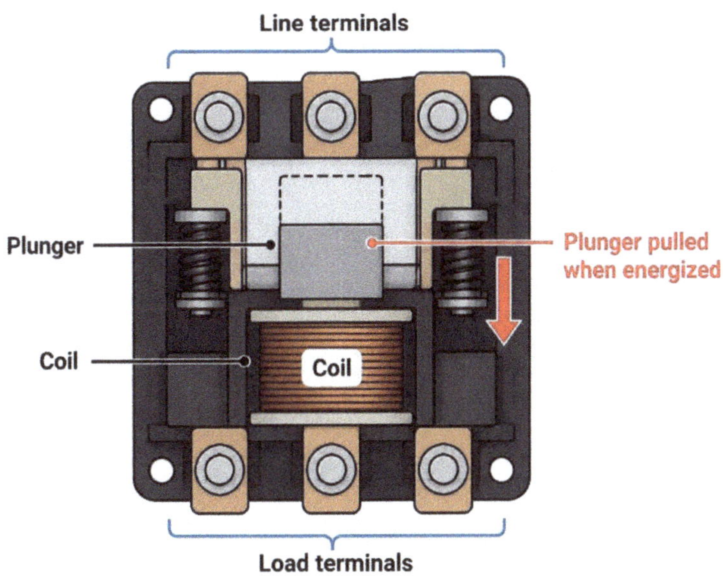

The Diagnostic Mindset: Notice how the entire outdoor system hinges on **Step 4**. If you go outside and see that the contactor plunger is physically pulled in, but neither the compressor nor the fan is running, you instantly know your 24V control system is perfectly healthy—your problem lies entirely within the 240V high-voltage circuit or the dual-run capacitor.

3.4 Why Cooling Mode Fails (The Big Three)

1. Electrical Failure (Most Common)

- Bad capacitor
- Bad contactor
- No 24V signal
- Blown fuse
- Bad transformer

2. Airflow Failure

- Dirty filter
- Dirty evaporator coil
- Weak blower motor
- Blocked return vents

3. Refrigerant-Side Issues (Diagnosis Only)

- Low charge
- Restrictions
- Failed compressor

You will learn how to diagnose these safely and legally — without touching refrigerant.

3.5 What You'll Learn Next

Now that you understand both heating and cooling sequences, the next chapter teaches the **low-voltage control system** — the wiring that makes everything talk to each other.

CHAPTER 4 — Low-Voltage Control Systems (24V Thermostat Wiring)

The Nerve System of Every Furnace and Air Conditioner

If the sequence of operation is the **brain**, then the low-voltage wiring system is the **nervous system** of HVAC.
It carries the signals that tell the furnace, blower, and outdoor AC unit what to do — and when to do it.

Most HVAC failures that look "mysterious" are actually simple low-voltage issues:

- A broken thermostat wire
- A blown 3-amp fuse
- A loose connection at the control board
- A missing C-wire
- A short between R and C
- A thermostat miswired by the previous homeowner

This chapter demystifies the entire 24-volt control system so you can diagnose these problems with confidence.

4.1 What the 24-Volt System Actually Does

Your furnace has a small transformer that converts 120V household power into **24 volts AC**.

This 24V power:

- Feeds the thermostat
- Powers the control board
- Energizes relays
- Pulls in the outdoor contactor
- Activates gas valves
- Closes circuits for heat, cool, and fan

It's low voltage, but it's the **command system** for the entire HVAC unit.

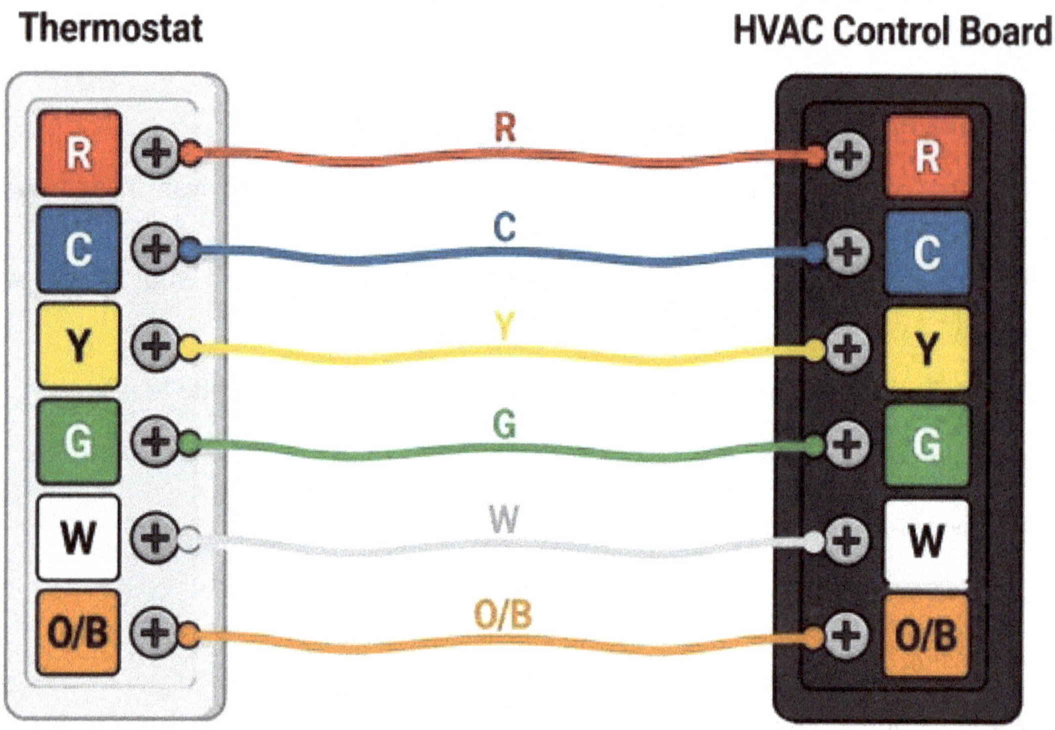

4.2 The 24V Transformer (The Heart of the Control System)

Every furnace contains a transformer that outputs:

- **R (hot)** — 24V power
- **C (common)** — the return path

Together, R and C form the **24V power loop**.

If the transformer fails, the entire HVAC system becomes "brain-dead."

Symptoms of a failed transformer:

- Thermostat is blank
- No 24V between R and C
- No LED lights on control board
- 3-amp fuse may be blown

Transformers fail most often due to shorts in thermostat wiring.

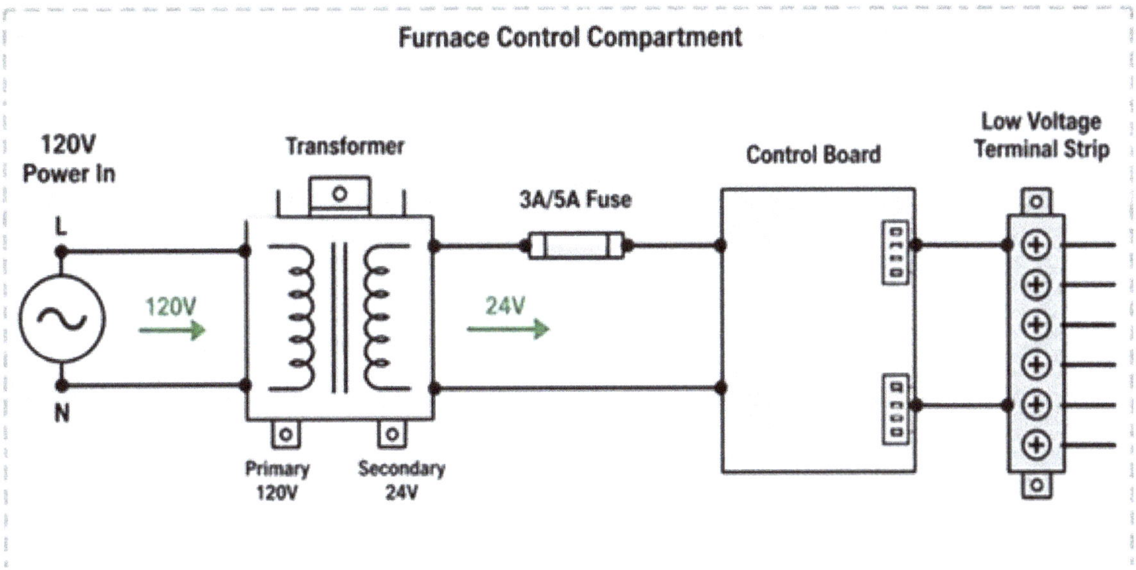

4.3 Standard Thermostat Wire Colors (And Why You Can't Trust Them)

In a perfect world, every HVAC installation would follow the exact same color-coded roadmap. The industry *does* have a standardized color convention for low-voltage control wiring. If a seasoned professional wired your home, the color of the insulation jacket will match the letter stamped on the terminal block.

However, as a diagnostician, you must internalize one absolute truth: The control board and the thermostat do not care about the color of the plastic insulation. Electricity only cares about the continuous copper path.

The Industry-Standard Color Map vs. Field Reality

Terminal	Standard Color	Primary System Function	Real-World Reality Check (What You Actually Find)
R	Red	24 VAC Hot Power (From the transformer)	This is almost always Red. However, in older systems with separate heating and cooling transformers, you may see an Rc (Cooling power) and an Rh

Terminal	Standard Color	Primary System Function	Real-World Reality Check (What You Actually Find)
			(Heating power) split by a metal jumper clip.
C	Blue or Black	24 VAC Common (The return path)	The Most Dangerous Wire. Just because a blue wire is tucked behind your thermostat does *not* mean it is connected to Common at the furnace board. Unlicensed installers often leave it disconnected at the furnace end, leaving smart thermostats unpowered.
W	White	Heating Call (Engages gas valve/heat sequence)	Typically reliable as the heat call wire in standard split systems. In heat pump applications, this wire might bridge to Aux/E to control backup electric heat strips.
Y	Yellow	Cooling Call (Engages compressor/outdoor fan)	Often yellow, but if an installer ran out of standard 5-wire thermostat cable, they might use a blue, brown, or orange wire to carry the outdoor cooling call.
G	Green	Indoor Fan Continuity (Blower motor relay)	Commands the indoor fan to turn on independently when you switch the thermostat toggle from "Auto" to "On".

Terminal	Standard Color	Primary System Function	Real-World Reality Check (What You Actually Find)
O / B	Orange (or Black)	Reversing Valve (Heat Pump systems only)	Critical for heat pumps. An O signal typically energizes the reversing valve in cooling mode (standard for Rheem/Ruud), while a B signal energizes it in heating mode (standard for Bosch/Trane).
Aux / E	Brown	Auxiliary / Emergency Heat	Activates the secondary electric resistance heat strips inside an air handler when the outdoor heat pump can no longer keep up with extreme winter temperatures.

The Technician's Golden Rule: Verify, Don't Assume

Homeowners, general handymen, and rushed installers frequently use whatever wire bundle is clutched in their truck box. If they ran short on a Friday afternoon, they might use a green wire for cooling and a yellow wire for the fan.

If you wire a new smart thermostat based strictly on a color chart without verifying where those wires actually land on the furnace terminal strip, you risk causing an instantaneous direct short circuit.

The Line-of-Sight Mandate: Before disconnecting a single wire from your old thermostat, walk down to the basement or utility closet, remove the lower door panel of your furnace, and look at the control board terminal strip. If the white wire is screwed into the W terminal at the board, then it is a heating wire. If the blue wire is screwed into the Y terminal at the board, then blue is your cooling call. The control board terminal letter always dictates the absolute truth, never the wire color.

Testing to Find an "Unknown" Wire Color

If you are dealing with an old, unlabelled cable bundle where all the wires look faded or dusty, use your multimeter to find the real identity of the wires:

1. Turn off the main electrical breaker to the furnace.
2. Go down to the furnace control board and temporarily twist the R (Red) and W (White) wires together.
3. Return to the wall thermostat upstairs, set your meter to Continuity (\rightarrow|- / \Omega).
4. Touch your probes across the pairs until your meter beeps. When it beeps, you have successfully proven that specific pair forms the continuous loop to the **heating circuit, regardless of what color the insulation is.**

4.4 How the Thermostat Communicates With the Furnace

The thermostat is nothing more than a **switchboard**.

When it wants something to happen, it closes a circuit:

- **Heat:** W → R
- **Cool:** Y → R
- **Fan:** G → R

The furnace control board sees these signals and responds accordingly.

Example: Cooling Mode

1. Thermostat closes Y → R
2. 24V travels to the control board
3. Board sends 24V outside to the contactor
4. Contactor pulls in
5. Compressor and condenser fan start

This is why a broken Y wire means **no cooling**.

4.5 The C-Wire (Common Wire) — The Most Misunderstood Wire in HVAC

The **C-wire** completes the 24V circuit and provides constant power to:

- Smart thermostats
- Wi-Fi thermostats
- Digital thermostats

If your thermostat doesn't have a C-wire, you may experience:

- Thermostat rebooting
- Thermostat losing Wi-Fi
- Furnace short cycling
- Transformer overload

Solutions for missing C-wire:

- Add-a-wire kit
- Run a new cable
- Use a plug-in transformer
- Use a thermostat that doesn't require C

You'll learn these in detail in Chapter 13.

4.6 The Furnace Control Board (The Command Center)

The control board receives thermostat signals and decides:

- When to start the inducer
- When to energize the igniter
- When to open the gas valve
- When to start the blower
- When to send 24V outside to the AC contactor

It also monitors:

- Limit switches
- Pressure switches
- Flame sensor
- Rollout switches

And it reports problems using **LED flash codes**.

Common flash codes:

- 1 flash: Ignition failure
- 2 flashes: Pressure switch stuck open
- 3 flashes: Pressure switch stuck closed
- 4 flashes: High-limit switch open
- Steady on: Normal operation

Every brand uses its own pattern, but the logic is the same.

4.7 The Low-Voltage Signal Mapping Matrix

To troubleshoot a low-voltage issue, you must look at the system as a communications network. The thermostat acts as the transmitter, the furnace control board acts as the central router, and the individual components act as the receivers.

When a specific mode is activated, 24 VAC power leaves the transformer and travels along a precise pathway. If that pathway is interrupted, the system goes deaf to the thermostat's commands.

The Master Low-Voltage Path

Use the data layout below to track exactly how 24V power routes through the entire system across all operational modes:

```
 [ 24V Transformer ] ---> (R Terminal) ---> [ Thermostat Switchboard ]

                              |

    +-----------------+-------------------------+-----------------+

    | (Heat Mode)     | (Cool Mode)                  | (Fan Only)

    v                 v                              v

  [ W Wire ]      [ Y Wire ]                    [ G Wire ]

    |                 |                              |

    v                 +---------------------+        v

 [ Furnace Board ]    v                    v      [ Furnace Board ]

    |           [ Furnace Board ]   [ Outdoor Contactor ]     |

    v                   |                  |                  v

 - Inducer Relay        v                  v          - Indoor Blower

 - Ignition Control  - Indoor Blower   - Compressor Start     Relay (Low Speed)

 - Gas Valve Relay   Relay (High Speed)   - Condenser Fan Start
```

Control Terminal Reference Guide

Terminal / Wire	Standard Color	Signal Path & Target Destination	What Happens When Energized
R	Red	From: 24V Transformer Primary To: Thermostat R Terminal	Provides continuous 24 VAC hot power to constant-memory thermostats and the switchboard.
C	Blue / Black	From: Furnace Board Common To: Thermostat C Terminal	Completes the electrical circuit path back to the transformer to power digital displays and Wi-Fi radios.
W	White	From: Thermostat Heating Contacts To: Furnace Board W Terminal	Initiates the gas heating sequence (Inducer motor start, safety checks, and ignition).
Y	Yellow	From: Thermostat Cooling Contacts	Splices at the indoor board to simultaneously start the high-speed indoor blower and energize the outdoor contactor coil.

Terminal / Wire	Standard Color	Signal Path & Target Destination	What Happens When Energized
		To: Furnace Board Y & Outdoor Contactor	
G	Green	From: Thermostat Fan Switch To: Furnace Board G Terminal	Commands the indoor blower motor to run independently on a lower continuous speed for air filtration.

The Line-of-Sight Rule: When you remove the lower door panel of a furnace, you will find a low-voltage terminal strip matching these exact letters (R, C, W, Y, G). If you ever find yourself confused by a system failure, you can skip the thermostat completely by safely jumping R to W (for heat) or R to Y (for cooling) directly on the board with a small piece of insulated wire. If the system starts, you have instantly proven the furnace is healthy, and the problem lies entirely in the thermostat or the wall wiring.

4.8 Common Low-Voltage Failures (And How to Diagnose Them)

Low-voltage failures are the ultimate tricksters of residential HVAC. Because a 24 VAC circuit relies on thin, fragile wiring and sensitive magnetic coils, a failure here can easily mimic a major system catastrophe.

Before you assume a compressor is dead or a furnace heat exchanger is ruined, work through these six common low-voltage failure modes with your multimeter.

1. The Blown 3-Amp or 5-Amp Board Fuse

The small plastic automotive-style fuse on your furnace control board is the sacrificial lamb of the low-voltage system. It is designed to pop instantly if too much current flows through the circuit, protecting your expensive control board and transformer from burning up.

- The Primary Cause: A direct electrical short circuit between R (24V Hot) and C (Common), or between R and the bare metal chassis ground. This frequently happens when thin thermostat insulation rubs raw against sharp sheet-metal cabinet edges during fan vibrations.
- The Diagnostic Routine: 1. Turn off the furnace power breaker.

2. Locate the brightly colored fuse (usually purple for 3-amp, amber for 5-amp) on the face of the control board.

3. Pull it straight out. Hold it up to a light source. If the curved metal filament inside the clear plastic housing is broken or the plastic is scorched, it is blown.

4. *Do not just replace it.* If you insert a new fuse without finding the short, it will pop the second you turn the power back on. Inspect the thermostat bundle for pinch points or raw copper before replacing.

2. Broken Thermostat Wire

Thermostat wires are typically thin 18-gauge solid copper. They are brittle and highly susceptible to physical damage, especially where they run through unfinished basements, crawlspaces, or outdoors to the AC condenser.

- The Primary Cause: Pest damage (mice chewing through insulation), physical impact from weed-whackers on the outdoor run, or dry-rot from sun exposure.
- The Diagnostic Routine: * If the system suffers from *no heat* but the fan runs manually, set your meter to AC Voltage.
    - With a call for heat active at the thermostat, place your probes on W and C at the furnace board.
    - If you read 0 VAC at the board but the thermostat upstairs says it is heating, the W wire is physically severed somewhere inside the walls. Repeat this test between Y and C for cooling failures.

3. The Miswired Thermostat

This failure mode almost always occurs right after a homeowner attempts to install a new digital or smart thermostat themselves without taking a "before" photograph of the old wiring layout.

- The Symptoms: The indoor blower fan refuses to run, the outdoor AC turns on when the system is switched to heating mode, or the furnace cycles but never attempts ignition.
- The Diagnostic Routine: * Kill the power, pop the thermostat off its wall baseplate, and check the terminal letters.
  - Cross-reference them with the terminal letters on your furnace control board strip. If the green wire is in G at the furnace but shoved into C or Y at the thermostat, the system logic is completely crossed. Re-terminate the wires to match the exact terminal mapping established at the furnace board (revisit Section 4.3).

4. Bad 24V Transformer

The transformer is the powerhouse of the control system. It takes 120V incoming household power and drops it down to 24V. If the transformer dies, the entire machine loses its mind.

- The Symptoms: The thermostat screen is completely blank, there are no diagnostic LED flash lights visible on the control board, and the system is dead silent.
- The Diagnostic Routine:
  1. Verify the furnace has 120V incoming power at the main door switch.
  2. Set your multimeter to AC Voltage.
  3. Touch your probes to the two output screw terminals of the transformer (usually labeled SEC or connected to the black and blue low-voltage wires).
  4. The Verdict: If your meter reads a flat 0.0 VAC, the transformer's internal primary or secondary windings have melted down due to an unresolved short circuit. Replace the transformer.

5. Failed Contactor Coil

The outdoor contactor uses a small magnetic coil to pull down its heavy line-voltage contacts. If this low-voltage coil burns out, high-voltage power will never reach your compressor or outdoor fan.

- The Symptoms: The indoor system runs beautifully, but the outdoor unit remains dead silent. When cooling is called, you never hear the distinct mechanical "click" or heavy "clack" sound of the outdoor contactor pulling in.
- The Diagnostic Routine:
    1. Go to the outdoor service panel. Set your meter to AC Voltage.
    2. Place your meter leads directly on the two small side brass terminals of the contactor coil body.
    3. The Verdict: If your meter reads 24 VAC, the thermostat signal successfully made it outside. If that 24V is present but the metal plunger isn't pulled down, the contactor coil is mechanically seized or electrically open. Replace the contactor.

6. Loose Terminal Connections

Low-voltage wires are held in place by basic screw-down clamping blocks. Over time, seasonal temperature swings cause copper wires to expand and contract, which can slowly back the terminal screws out.

- The Symptoms: The system shuts down randomly mid-cycle, short-cycles rapidly, or provides intermittent cooling where it works perfectly one day and fails the next.
- The Diagnostic Routine:
    - Turn off all system power.
    - Take a small precision cabinet screwdriver and manually check the tightness of every single low-voltage wire screw at the thermostat baseplate, the furnace control board, and the wire-nut wire splices leading out to the AC unit.
    - If a wire can wiggle or spin under its screw head, strip it back to fresh copper, re-seat it, and torque the terminal screw down snug.

TECH ALERT: The Short-Circuit Trap: If you discover a blown 3-amp board fuse, never put a larger 5-amp or 10-amp fuse in its place just to keep the system running. Forcing a larger fuse into a low-voltage circuit with an active short will instantly bypass the safety protection, melting the low-voltage wires together inside your walls or burning the furnace control board to a crisp. Always locate the short circuit first!

4.9 What You'll Learn Next

Now that you understand the low-voltage system, you're ready for the most powerful section of the entire book:

CHAPTER 5 — The Multimeter in HVAC Work

Your Most Important Tool for Diagnosing Heat, Cool, and Airflow Problems

If you want to diagnose HVAC systems like a technician, you must master one tool above all others: the multimeter.

You do not need an expensive or overly complex model. A reliable, rugged trade meter just needs the capability to read four critical metrics:

- 24 VAC (Low-voltage control paths)
- 120/240 VAC (High-voltage lines)
- Continuity (Circuit paths and switches)
- Microfarads (μF) (Motor run capacitors)

This chapter teaches you exactly how technicians use a multimeter in the field—safely, confidently, and without guesswork. Mastering this tool allows you to isolate faults across transformers, thermostat circuits, control boards, contactors, capacitors, limit switches, pressure switches, and flame sensors. This is the core knowledge that saves operators and homeowners hundreds of dollars in avoidable service calls every season.

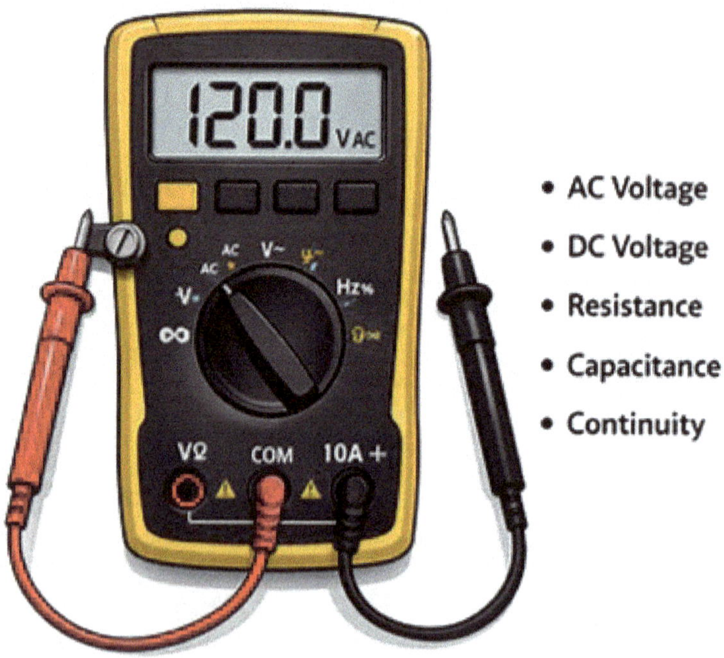

5.1 Safety First — The Rules of Meter Use in HVAC

Before touching a meter probe to a single terminal or wire, commit these six fundamental laws to memory. Working with electricity requires absolute focus—there are no shortcuts. These rules do not just protect your diagnostic tools; they keep you alive.

Rule 1 — Treat Every Wire as Live

Never assume a circuit is dead just because a switch is turned off or a breaker is flipped. Field wiring can be unpredictable, and double-fed circuits are a common hazard. Always use your meter to verify there is absolutely zero voltage present before touching a bare wire or terminal with your hands.

Rule 2 — Keep Your Fingers Behind the Probe Guards

Multimeter probes are engineered with raised plastic ridges near the tips. These are safety finger guards. Always grip the plastic probe handles firmly behind these ridges. Never let your fingers slip forward to touch the exposed metal tips while testing a live circuit.

Rule 3 — Prove Your Meter Before and After Testing

Professionals use the Live-Dead-Live protocol (also called meter proving). Before testing a circuit to see if it is safe to touch, place your probes on a source you know is live, such as a standard wall outlet or incoming line power. If the meter reads correctly, you have proved the tool works. Test your target circuit. Then, check the known live source one more time to verify the meter did not fail or blow an internal fuse during the measurement.

Rule 4 — Set the Meter Function Before Touching the Circuit

Never change the dial settings on your multimeter while the probe tips are actively touching a live circuit. Arcing can occur inside the selector switch, destroying the tool or causing an electrical short. Select your specific setting (such as AC Voltage), verify the leads are plugged into the correct ports on the meter body, and then take your reading.

Rule 5 — Never Test Resistance (Ohms) or Continuity on a Live Circuit

The resistance and continuity settings on a multimeter work by sending out a tiny, delicate electrical current from the meter's own internal battery to check how easily it flows through a component. If you introduce outside line voltage (120V or 240V) into the meter while it is in resistance mode, you will instantly fry the meter's internal circuitry and potentially cause the tool to fail dangerously in your hand. Always isolate the component and pull the wires off it before testing resistance.

Rule 6 — Maintain Total Isolation When Testing High Voltage (240V)

Old-school instructional manuals often advise keeping one hand behind your back to prevent current from traveling across your chest if you get shocked. In reality, testing a two-terminal contactor or disconnect box requires using both hands to steady the meter leads safely on the contacts. Instead, practice Total Isolation:

- Ensure your hands are completely dry and you are wearing rubber-soled work shoes.
- Never lean your body, shoulders, or bare elbows against the metal cabinet frame of the HVAC unit while testing live power.
- If you are uncomfortable manual-testing a high-voltage system, invest in a pair of insulated alligator clip leads. Turn the power off at the breaker, safely clip the leads onto the target terminals, step back, and flip the power switch on to read the measurement hands-free.

The Golden Rule: Meters do not lie, but human assumptions do. If your meter tells you a wire is live—even if you are positive you turned the breaker off—trust the meter every single time.

5.2 Understanding the Three Voltages in HVAC

Residential equipment relies on a clear division of three distinct electrical levels:

1. 24 VAC (Low Voltage): Functions as the safety and command brain. It powers the thermostat, contactor coils, gas valve solenoids, and circuit board logic loops.
2. 120 VAC (Line Voltage): Functions as the localized power muscle. It drives the indoor furnace blower motor, the combustion draft inducer, the hot surface igniter, and primary control board transformers.
3. 240 VAC (High Voltage): Functions as the heavy machinery muscle. It supplies the raw electrical power required to drive the outdoor air conditioner compressor pump and the condenser fan motor.

5.3 How to Test for 24 VAC (The Control Circuit)

Set your multimeter dial to AC Voltage (V\sim or VAC).

To track down a low-voltage failure or trace a thermostat command, keep your black meter probe grounded to the C (Common) terminal on the main furnace board strip, and use your red probe to check the active command terminals:

- R to C: Should read a constant 24V AC to 28V AC. This proves the step-down transformer is outputting power.
- R to W: Measures the heating call path.
- R to Y: Measures the cooling call path.
- R to G: Measures the indoor fan circulation path.

Diagnostic Evaluation:

- If R to C reads 0V: You have a bad transformer, a blown 3-amp or 5-amp automotive blade fuse on the board face, a tripped high-voltage breaker feeding the indoor unit, or a severed low-voltage line.
- If R to Y reads 0V during an active cooling call at the wall: The thermostat is failing to close its internal relays, the thermostat sub-base is broken, or a wire is broken inside the wall jacket between the living space and the furnace cabinet.

5.4 How to Test for 120 VAC (Furnace Line Voltage)

Set your multimeter to AC Voltage. Ensure your meter leads are plugged into the standard voltage ports. When an indoor component fails to run, this test determines the ultimate diagnostic answer: Is the part itself dead, or is the control board failing to send it power?

Where to Test for 120 VAC

- Incoming Power Lugs: Where the main household power feed enters the furnace junction box.
- Inducer Motor Leads: At the wiring plug connecting the inducer assembly to the control board terminal.
- Blower Motor Terminals: Between the active speed taps (labeled HEAT or COOL) on the circuit board and the neutral terminal block.
- Hot Surface Igniter Plug: At the plastic quick-connect harness leading directly to the igniter element.

Diagnostic Matrix: Reading the Results

Scenario A: The Furnace Has No 120V Power At All

If you test across the incoming black (hot) and white (neutral) lines entering the cabinet junction box and get a reading of 0 VAC, the issue is completely external to the furnace electronics. Check for a tripped circuit breaker in the main home panel, a turned-off service switch on the side of the furnace cabinet wrapper, or a loose wire nut inside the connection box.

Scenario B: The Inducer Motor Won't Spin

Wait for the thermostat to call for heat. Test across the two wires leading directly into the inducer motor plug.

- If you read 120 VAC but the motor remains silent and cold: The control board is working perfectly. The inducer motor winding is burned open or the internal bearings are mechanically seized. Replace the inducer motor assembly.
- If you read 0 VAC: The control board is failing to close its internal line-voltage relay. Replace the control board.

Scenario C: The Blower Motor Won't Spin

Wait for the furnace to initiate ignition and reach its designated blower-on delay time (typically 30 to 45 seconds). Test between the active blower terminal speed tap on the board (HEAT or COOL) and the white Neutral terminal block.

- If you read 120 VAC but the blower fan is silent: The board is delivering proper power. The blower motor winding is dead, or its run capacitor has dropped below operational tolerance.
- If you read 0 VAC: The board's internal blower relay has failed to close. Replace the control board.

Scenario D: The Hot Surface Igniter Won't Glow

Wait for the sequence of operation to reach the ignition phase. Test across the two pins inside the wire harness plug coming from the control board to the igniter.

CRITICAL VOLTAGE WARNING: Before condemning a part, look at the manufacturer label on the furnace cabinet wrapper. Traditional silicon carbide igniters require 120 VAC from the board. However, modern high-efficiency units often use rugged silicon nitride igniters engineered to operate on 24 VAC or 80 VAC.

- If you read the correct designated operating voltage on your meter face but the igniter remains dark: The board is healthy. The igniter element is cracked or broken. Replace the igniter.
- If you read 0 VAC during the ignition step: The control board is not routing power down the line. Replace the control board.

5.5 How to Test for 240 VAC (Outdoor High Voltage)

Testing for 240 VAC is the definitive diagnostic path when the outdoor condenser unit remains completely silent during an active call for cooling. This test traces high-voltage electrical power from your home's main breaker box down to the outdoor unit's core components.

Set your multimeter dial to AC Voltage (V\sim). Ensure your meter is rated for at least 600V.

The 4-Point High-Voltage Check

Always check for voltage in a logical sequence, moving from the house power source outward toward the machine:

1. The Incoming Disconnect Lugs (Line Side): Open your outdoor disconnect box mounted to the house wall and place your meter probes on the two main incoming power terminals. You should read between 220V and 240V AC. If you read 0V, look for a tripped or faulty circuit breaker in the main home panel, or a broken underground feeder line.
2. The Disconnect Output (Load Side): Insert the disconnect pull-out plug (or turn the mechanical switch to ON) and test the output terminals. If you read 0V here but had power on the line side, the disconnect pull-out plug has a blown cartridge fuse, or the internal disconnect switch mechanism has failed.
3. The Contactor Line Terminals: Move down into the electrical service panel of the AC unit itself. Place your probes on the bottom two terminals of the contactor (labeled LINE). This proves power is successfully traveling through the outdoor whip conduit.
4. The Contactor Load Terminals: Place your probes on the top two terminals of the contactor (labeled LOAD) while the contactor plunger is physically depressed.

5.6 How to Test a Contactor (Cooling Mode)

The contactor is an electromechanical switch. When the indoor control board sends a 24V signal to the contactor's internal magnetic coil, it pulls down a heavy metal bridge, allowing 240V power to pass across the contacts to fire up the compressor and fan.

The Three-Step Contactor Test Protocol

Test 1: Coil Voltage (The Command Signal)

Set your meter to AC Voltage. Place your probes on the two small brass terminals located on the left and right sides of the contactor body frame.

- The Target: When the thermostat is actively calling for cooling, you must read between 24V and 28V AC.
- The Diagnosis: If you read 24V but the contactor plunger refuses to pull in magnetically, the internal coil windings are burned out or the mechanism is physically seized. Replace the contactor. If you read 0V, the outdoor unit isn't receiving the command from the indoor furnace board.

Test 2: Line Voltage (The Supply Power)

Keep your meter on AC Voltage. Place your probes on the two bottom LINE terminals.

- The Target: You must read a steady 240V AC at all times, regardless of whether the system is running or idle.

Test 3: Load Voltage (The Delivery Power)

With the contactor plunger pulled in (either magnetically by a call for cooling, or carefully pushed in using an insulated tool for a brief test), place your probes on the two top LOAD terminals.

- The Target: You should read 240V AC.
- The Diagnosis: If you verify 240V on the LINE side and the contactor pulls in, but you read 0V on the LOAD side, the copper contact points inside the switch are severely pitted, burned, or blocked by bugs (ants are notoriously attracted to live contactor points). The switch cannot conduct electricity. Replace the contactor.

5.7 How to Test a Capacitor (The μF / MFD Test)

The dual-run capacitor is the single most common failure point in residential air conditioning systems. It acts as an electrical booster, storing energy to keep the compressor and condenser fan motor running efficiently. When a capacitor weakens or dies, the compressor will hum loudly but fail to spin, eventually overheating and shutting down.

Step-By-Step Bench Testing Protocol

To get a technically accurate reading, you must completely isolate the capacitor from the system circuit.

1. Safely De-Energize: Turn off the outdoor disconnect power and safely discharge the capacitor using an insulated screwdriver across the terminals.
2. Label and Remove Wires: Take a quick photo of the wiring layout with your phone. Disconnect the wires from the three terminal blocks using needle-nose pliers.
3. Set the Meter: Turn your multimeter dial to the Capacitance setting, denoted by the -||- or μF symbol.
4. Test the Compressor Circuit: Place the black meter probe on the C (Common) terminal and the red probe on the HERM (Hermetic / Compressor) terminal. Wait 2 to 3 seconds for the digital display to stabilize.
5. Test the Fan Circuit: Leave the black probe on the C terminal and move the red probe to the FAN terminal. Note the reading.

Calculating the Manufacturer Tolerance Range

Every capacitor has its microfarad (μF) ratings stamped on the side of the metal casing, followed by a tolerance limit—almost always ±6%. If your meter reading drops below that 6% window, the capacitor must be replaced. To calculate your specific capacitor's passing range, multiply the rating by 0.94 (the bottom limit).

Real-World Example: Reading a 45 / 5 μF Capacitor

- Testing the HERM Circuit (Rated for 45 μF): $45 \times 0.94 = 42.3\text{ μF}$ (Minimum Acceptable Reading). If your meter reads anywhere below 42.3 μF, the compressor will struggle to start or run hot.
- Testing the FAN Circuit (Rated for 5 μF): $5 \times 0.94 = 4.7\text{ μF}$ (Minimum Acceptable Reading). If your meter reads below 4.7 μF, the outdoor fan motor will run slowly or overheat.

Diagnostic Rule of Thumb: If your meter reads a flat 0.0 μF on either circuit, the internal layers of aluminum film and oil have completely ruptured. The part is entirely dead.

5.8 How to Test Limit Switches (Heat Mode)

Limit switches are safety devices designed to protect your home from fire and your furnace from structural damage. They are Normally Closed (NC) switches. Under normal conditions, they form a solid, unbroken path for electricity. If the temperature inside the heat exchanger cabinet rises to a dangerous level, the internal bi-metal disc snaps open, breaking the control circuit and shutting down the gas valve.

Testing for Continuity

1. Turn off all electrical power to the furnace.
2. Disconnect at least one of the two slip-on wire leads from the limit switch terminals to prevent reading a false circuit back-feed through the control board.
3. Set your multimeter to Continuity (the audio beep symbol) or Resistance (Ω).
4. Touch your meter probes to the two metal male terminals of the limit switch.

- The Result at Room Temperature: Your meter must beep instantly or display an ohm reading of 0.0 Ω to 0.3 Ω. If the meter remains silent or displays OL (Open Loop) while the furnace is cold, the switch is permanently broken and must be replaced.
- The Result During Operation: If the furnace starts up, fires normally, but then snaps its burners off after a few minutes while the indoor fan is running, perform the drop-down voltage test detailed in Section 5.4. If the switch reads 24V AC across its terminals while the burners are running, it has opened.

The Diagnostic Mindset: Limit switches are incredibly simple and rarely fail on their own. If a limit switch trips open, it is almost always doing its job. Do not simply change the switch. Look for a severe airflow restriction, such as a completely choked air filter, blocked return grills, or a failing indoor blower motor.

5.9 How to Test Pressure Switches (Heat Mode)

A pressure switch is a safety device designed to verify that the furnace exhaust gases are being safely vented out of the home. It is a Normally Open (NO) switch. It requires a specific amount of negative pressure (suction) from the combustion inducer fan to pull its internal diaphragm shut.

The Operational Testing Protocol

Because this switch changes state during operation, you must test it dynamically while the furnace attempts to run.

1. Set your multimeter dial to AC Voltage (V\sim).
2. Turn on the furnace power and set the thermostat to call for heat.
3. Keep your black meter probe securely grounded to the bare metal chassis of the furnace.
4. Take your red probe and touch the Input Terminal of the pressure switch (the wire coming directly from the control board transformer circuit). Your meter should read 24V AC.
5. Once the inducer motor reaches full operating speed, touch your red probe to the Output Terminal of the pressure switch.

Interpreting the Voltage Path

- Scenario A: 24V on both terminals. The inducer fan is pulling an adequate vacuum, the draft path is perfectly clear, and the switch has closed successfully. The safety loop is verified.
- Scenario B: 24V on input, but 0V on output. The inducer fan is running, but the pressure switch remains open. The control board will hang up at this step indefinitely, and the igniter will never glow.

Troubleshooting an Open Pressure Switch Circuit

If the switch refuses to close, do not assume the switch is bad. Work through the mechanical draft checklist before changing parts:

- Inspect the small clear vinyl or black rubber suction tube. Look for cracks, dry-rot, or water droplets trapped inside the line.
- Use a small paperclip to clear out the plastic nipple port on the inducer housing where the hose connects. This tiny hole frequently clogs with solid white zinc oxide or rust scale.
- Check the PVC venting termination outside the house for bird nests, ice accumulation, or leaf blockages.

5.10 How to Test Flame Sensors (The Flame Proving Test)

If the furnace burners ignite perfectly but snap completely off after exactly 3 to 5 seconds, you are dealing with a flame-proving failure. The safety logic on the control board thinks there is no fire, so it shuts off the gas valve to prevent unburned gas from filling your home. While a dirty sensor is the culprit 90% of the time, a technician doesn't guess—they test the signal directly using a multimeter.

The Core Concept: Flame Rectification

A flame sensor is just a stainless steel rod sitting directly in the fire. The control board sends an alternating current (AC) voltage out to the rod. When the burners light, the plasma inside the flame acts as a one-way gate, converting that AC voltage into a tiny direct current (DC) signal. This current flows through the flame, hits the metal burner assembly, and travels back to the control board through the furnace ground wire. This tiny signal is measured in Microamps DC (μA).

The 3-Step Diagnostic Protocol

Step 1: Visual Inspection (The Quick Check)

Turn off all power to the furnace. Locate the flame sensor (usually on the opposite side of the burner assembly from the hot surface igniter) and remove the single 1/4-inch screw holding it in place.

- What to look for: Look at the metal rod. Over time, burning gas deposits an invisible layer of clear silica (glass) or white oxidation over the steel. This acts as an electrical insulator, blocking the tiny microamp signal from reaching the flame.
- The Fix: Take a piece of light abrasive material (like a green Scotch-Brite pad or fine steel wool) and gently rub the metal rod until it is bright and clean. Never use sandpaper, as the flint/garnet on the paper will melt into a permanent glass coating the next time the burners light.

Step 2: The Ground Path Continuity Test

Because the microamp signal must travel through the flame and back to the board via the furnace frame, a loose or rusty ground wire will break the circuit just as easily as a dirty sensor.

- How to test: With the power off, set your multimeter to Continuity/Ohms (Ω). Place one probe on the metal burner bracket and the other on the main ground terminal or bare metal chassis of the furnace cabinet.
- The Result: Your meter should beep or read less than 1 Ω. If it doesn't, clean your ground connections and tighten the burner mounting screws.

Step 3: Direct Microamp (μA) Signal Test (Advanced)

If you have cleaned the sensor and verified the ground, but the furnace still drops out after 3 seconds, you need to measure the actual current.

1. Set your multimeter to Microamps DC (μA).
2. Unplug the single wire leading to the flame sensor.
3. Connect your meter leads in series: clip one probe to the loose wire coming from the board, and the other probe directly to the flame sensor terminal.
4. Turn the power back on and start a call for heat.
5. Watch the meter the exact moment the burners ignite.

Reading the Numbers: Most modern furnace control boards require a minimum signal of 1.0 to 1.5 μA to keep the gas valve open. A healthy, clean sensor typically reads between 2.0 and 5.0 μA. If your meter reads 0 μA or drops below the manufacturer's minimum rating during ignition, and the sensor is clean, the flame sensor rod or the control board itself is faulty.

5.11 What You'll Learn Next

Now that you know how to use a multimeter like a technician, you're ready for the next chapter: Chapter 6 — Advanced Testing and Replacing Run Capacitors.

CHAPTER 6 — Testing Capacitors (The #1 AC Failure)

The 15 Part That Causes 300 Service Calls Every Summer

If you have ever wondered why air conditioners fail most often on the hottest, most humid days of the year, here is the truth: capacitors are the single most common failure point in residential AC systems.

They are inexpensive, physically fragile, and constantly exposed to intense heat, mechanical vibration, and high electrical stress. When they fail, the entire outdoor cooling plant locks up and shuts down.

This chapter teaches you how to diagnose, remove, discharge, and bench-test capacitors safely and confidently—using nothing more than a standard trade digital multimeter equipped with a microfarad (\mu\text{F}) setting. Mastering this process allows you to preemptively catch weak, out-of-tolerance, or dead components before they trigger costly compressor or fan motor failures.

6.1 What a Capacitor Actually Does

An electrical capacitor functions as a heavy-duty, temporary storage booster canister for electrical energy. Single-phase alternating current motor windings cannot naturally establish the rotational magnetic field required to start spinning on their own. A run capacitor supports two critical outdoor motors:

1. The Compressor: The hardest-working, highest-amperage motor in the entire climate control system.
2. The Condenser Fan Motor: The top-discharge fan that pulls ambient outdoor air through the aluminum coil fins to reject heat.

Most residential air conditioning units utilize a dual-run capacitor, which houses two separate capacitor cores inside a single compact aluminum canister wrapper.

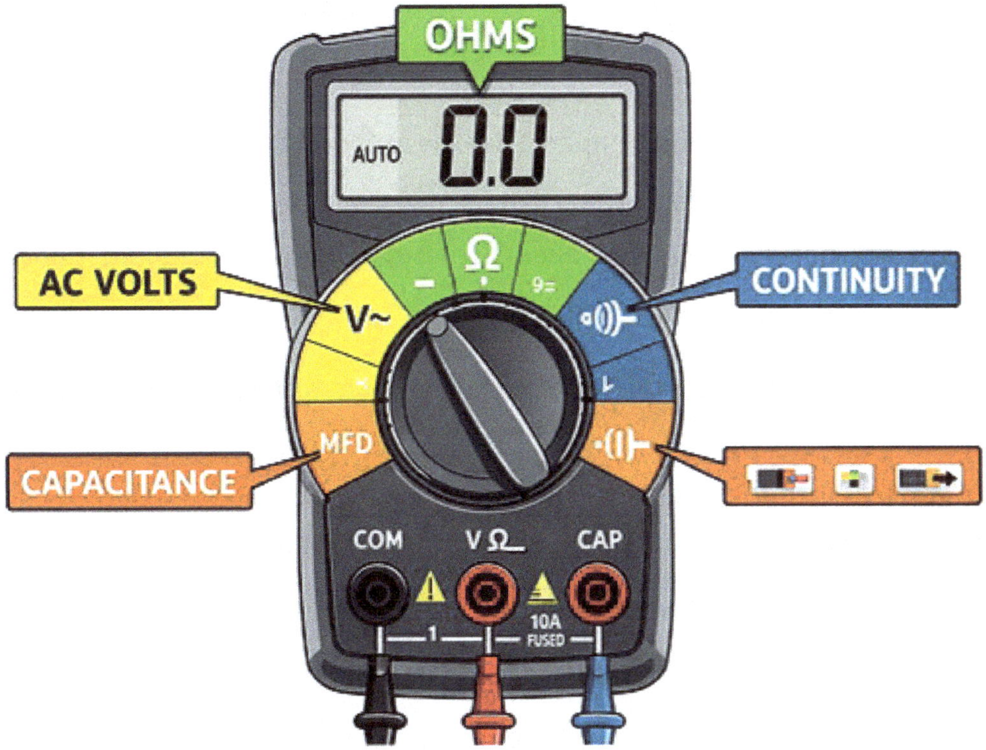

6.2 How to Identify a Dual-Run Capacitor

A standard dual-run capacitor is easily identified by the three distinct terminal clusters stamped onto its top plastic header plate:

- C (Common): The primary port where incoming line-voltage power feeds into the capacitor.
- HERM (Hermetic Compressor): The auxiliary output terminal wired straight to the compressor's start winding.
- FAN (Fan Motor): The auxiliary output terminal wired straight to the outdoor fan motor's start winding.

The manufacturer stamps the technical specifications directly onto the metallic casing skin. The label format typically reads:

40/5 µF 440V

This translates to:

- 40 µF: Dedicated capacitance capacity for the high-torque compressor circuit.
- 5 µF: Dedicated capacitance capacity for the outdoor fan motor circuit.
- 440 Volts: The maximum electrical voltage threshold the internal dielectric insulating oil and aluminum film sheets can safely withstand without rupturing.

6.3 Symptoms of a Bad Capacitor

Capacitors generally degrade gradually before failing catastrophically. Watch for these primary field warning signs:

- Outdoor Fan Refuses to Spin: The top-discharge fan remains stationary even during a call for cooling, though the motor housing emits a deep, hot electric hum.
- Compressor Hums But Fails to Start: The compressor attempts to draw locked-rotor current every few minutes, making a loud clicking and humming sound before tripping its internal thermal safety switch.
- AC Runs But Fails to Cool Effectively: A severely weakened capacitor drops the operational torque of the compressor, causing it to run under-speed and fail to compress the chemical refrigerant efficiently.
- System Cycles Off Prematurely: The compressor overheats rapidly under a weak phase shift and cuts out on its thermal overload loop after only a few minutes of run time.
- Physical Bulging or Oil Leaking: The structural top header plate or side casing of the aluminum can appears swollen, domed, or distorted.
- The High-Voltage Circuit Breaker Trips: The compressor draws extreme amperage trying to force its way past a dead start phase, instantly popping the home's main double-pole breaker.

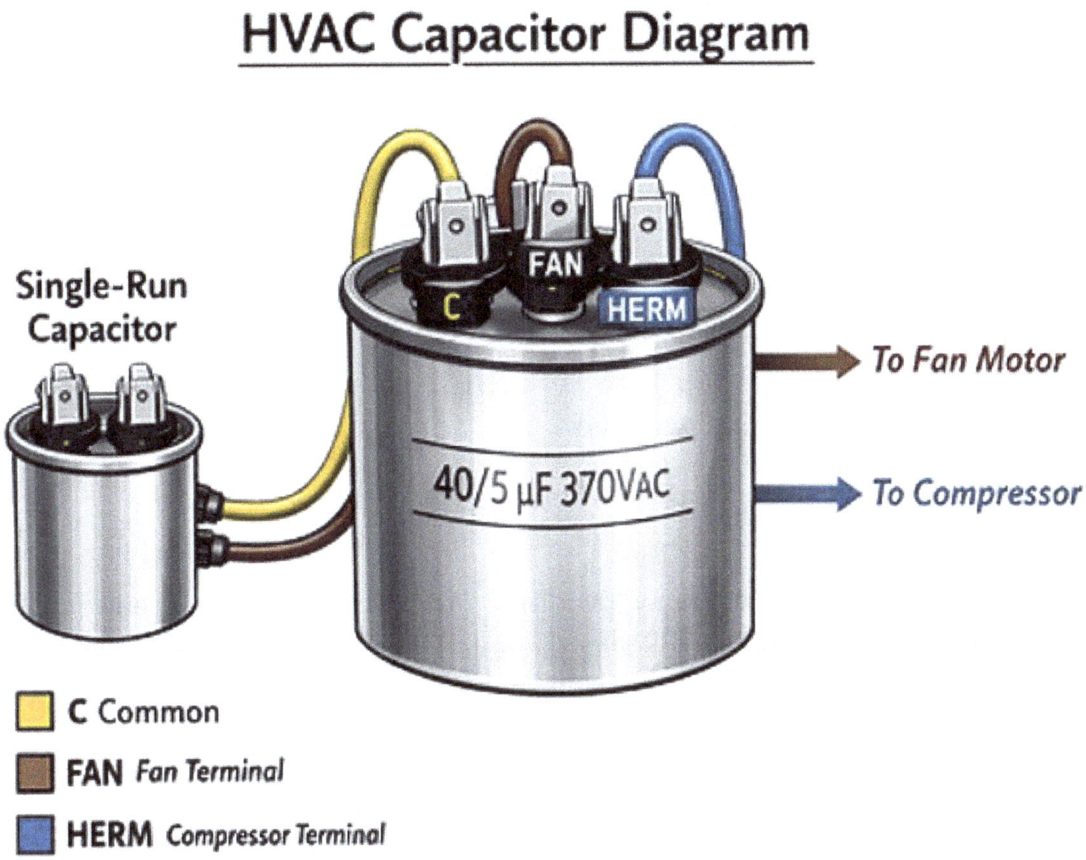

6.4 Safety First — How to Handle Capacitors

An air conditioner capacitor functions exactly like a high-voltage electrical storage cell. Even after you isolate the main circuit breaker and pull the outdoor electrical service disconnect block completely out of the wall cabinet, a healthy capacitor can retain its stored electrical charge for hours—or even days.

If you accidentally bridge these terminal clusters with your bare hands, it will deliver a violent, painful, and dangerous shock. Before putting a hand or a metal tool inside the electrical service panel wrapper, execute this non-negotiable safety protocol.

The 4-Step De-Energizing Protocol

Step 1: Kill the Power and Verify It's Dead

Never trust a simple visual check of a switch handle. Turn off the dedicated double-pole AC breaker inside the home's main electrical panel. Move outside to the condenser unit, open the metal service disconnect box mounted on the siding, and pull the internal plastic block completely out of the socket. Set your multimeter to AC Voltage. Test across the incoming line terminals on the lower half of the contactor relay to verify that absolute zero voltage is entering the cabinet wrapper.

Step 2: Discharge the Stored Energy Safely

Once line power is proven dead, you must bleed off the latent electrical charge hidden within the capacitor core layers. You can execute this step via two trade methods:

- Method A — The Insulated Screwdriver Trick (Field Standard): Take a heavy-duty screwdriver equipped with a solid, factory-rated insulated plastic or rubber handle. Grip the insulated handle firmly with one hand. Keep your other hand tucked cleanly behind your back or flat at your side. Press the bare metal shaft of the screwdriver flat across the terminals to form a solid bridge. Connect the C (Common) terminal cluster to the HERM terminal first, hold it steady for two seconds, and then pivot the metal shaft to bridge C to the FAN terminal cluster.

Field Note: If the capacitor was holding a peak charge, you may hear a sharp, loud "pop" and see a sudden electric spark jump to the tool shaft when the metal completes the gap. This is normal—it indicates the stored energy has been successfully neutralized.

- Method B — The Resistor Method (The Bench Standard): If you prefer to eliminate loud discharge snaps and protect your tool tips from microscopic pit burns, utilize a 20,000-ohm, 5-watt wire-wound resistor. Using insulated alligator clip test jumper leads, clamp one end of the resistor line to the C terminal, and touch the other lead to HERM, then repeat for FAN. The internal resistive matrix quietly bleeds off the electrical dump over 3 to 5 seconds, converting the energy into a tiny, imperceptible amount of thermal heat.

Step 3: Never Use Bare Metal Tools Without an Insulated Grip

Never grab a piece of uninsulated wire, bare metal pliers, or a pocket knife to short out a capacitor terminal. The current will always take the path of least physical resistance. If your hand tool lacks a dense, factory-rated rubber or plastic sleeve over the handle, that electrical path will travel directly through your body.

Step 4: Final Verification Test

Before loosening the mounting bracket or pulling wires off the terminals, use your multimeter to double-check your work. Set the meter dial to DC Voltage (V= or VDC), as stored capacitor energy behaves as direct current. Place your black probe on C and your red probe on HERM, then check FAN. The digital screen must display 0.0 V. If verified, the component is completely neutralized and safe to touch.

6.5 How to Remove and Discharge a Capacitor (Summary)

Follow this streamlined operational checklist whenever replacing a suspect unit:

1. Kill Power at the Disconnect: Pull the outdoor block completely out of the box.
2. Verify Power Is Off: Test for 240V AC at the bottom line side of the contactor.
3. Take a Photo of the Wiring: Document the factory wire coloring layout for your records.
4. Discharge the Capacitor: Bridge C to FAN and C to HERM using an insulated tool.
5. Remove Wires from Terminals: Pull the insulated spade connectors off using needle-nose pliers.

6.6 How to Test a Capacitor With a Multimeter (μF Test)

Set your digital multimeter dial to the Capacitance setting, denoted by the standard -||- or μF symbol.

The Bench Testing Sequence:

1. Isolate and safely remove the capacitor from its retaining strap inside the electrical service panel.
2. Ensure it has been completely discharged and all field wiring leads have been pulled off the spade terminals.
3. Place your black meter probe securely onto the C (Common) terminal cluster.
4. Touch your red meter probe to the HERM terminal cluster. Hold it perfectly still for 2 to 3 seconds until the digital screen stabilizes. Note the microfarad reading.
5. Keep the black probe on C, move your red meter probe over to the FAN terminal cluster, and record the second microfarad reading.

6.7 Understanding Tolerance (The ±6% Rule)

HVAC motor run capacitors are engineered to operate within a strict, narrow efficiency window. By industry convention, a capacitor must measure within ±6% of its factory-rated casing stamping to remain in service.

To quickly find the lower operating limit for any capacitor circuit, multiply the rated microfarad value by 0.94.

Real-World Analysis: Reading a 45 / 5 μF Casing Label

Evaluating the HERM Circuit (Rated for 45 μF):

$$45 \times 0.94 = 42.3\text{ μF (Minimum Allowable Field Reading)}$$

- If your multimeter reads 39.5 μF during your diagnostic check, the capacitor's internal film layers have degraded. It must be replaced immediately, even if the compressor is still managed to limp to a start.

Evaluating the FAN Circuit (Rated for 5 μF):

$$5 \times 0.94 = 4.7\text{ μF (Minimum Allowable Field Reading)}$$

- If your meter displays a value below 4.7 μF, the outdoor fan motor will suffer from low starting torque, run excessively hot, spin below designed RPM capacity, and eventually trip on its internal thermal overload.

Diagnostic Rules of Thumb:

- If your meter reads a flat 0.0 μF across the terminals, the internal film loops or terminal connection welds have snapped completely open. The part is entirely dead.

- If your digital display bounces erratically up and down without ever locking onto a stable number, the internal layers are shorting out across the dielectric fluid. The part is actively failing.

6.8 Visual Signs of a Bad Capacitor

You can frequently spot a failing capacitor during a basic visual inspection before you ever connect a diagnostic meter:

- Bulged or Domed Top Header Plate: The flat plastic or aluminum top surface has popped upward like a swollen can of food.
- Leaking Dielectric Oil: Clear or yellowish synthetic insulating oil pooling around the base or leaking from the terminal seams.
- Corroded or Rusted Terminals: Heavy orange scale or white oxidation blocking the copper spade connection points.
- Swollen, Out-of-Round Sides: The cylindrical aluminum casing appears warped or pushed out of its normal shape.
- Burn Marks or Scorched Metal: Black carbon arcs or melted plastic jackets around the wiring connectors.
- Loose, Wobbly Terminals: The internal rivets holding the spade heads to the core plates have melted or snapped free.

6.9 Why Capacitors Fail

Capacitors fail primarily due to five environmental and systemic factors:

1. Intense Thermal Heat: Outdoor condenser cabinets bake in direct summer sunlight, pushing internal electrical panel temperatures past 140°F (60°C), which degrades the internal insulating oil.
2. Mechanical Vibration: Constant harmonics generated by the heavy compressor pump shake the internal layers of aluminum film, causing physical wear over time.
3. Line Voltage Spikes: Power grid fluctuations, localized lightning strikes, or a failing indoor transformer can introduce severe voltage spikes that puncture the thin internal film sheets.
4. Operational Age: Standard residential run capacitors are wear items with an average engineered lifespan of 5 to 10 years.
5. Low-Quality Manufacturing: Many factory standard capacitors utilize thin, lower-grade components that struggle to withstand sustained peak summer loads.

6.10 The Technician's Shortcut: The Stick Test

If the outdoor AC unit is calling for cooling, the compressor is running, but the top fan blade remains completely stationary while humming hot, you can perform a rapid diagnostic verification known as the Stick Test:

Take a dry wooden stick, a rolled-up magazine, or an insulated screwdriver shaft. Slide it carefully through the top wire grille and give one of the fan blades a gentle, deliberate push in its normal direction of rotation. Clear the tool away from the grille instantly.

- The Result: If the fan blade picks up speed, spins up normally, and continues running on its own after your push, the run capacitor's FAN circuit is completely dead. * The Explanation: Your physical push substituted for the missing starting torque phase shift normally provided by the capacitor. This confirms the fan motor winding itself is healthy, but the electrical booster circuit is missing. Always confirm this shortcut with your multimeter before ordering parts.

6.11 Replacing a Capacitor (Safe & Compliance Metrics)

Replacing an external motor run capacitor is purely an electrical maintenance task. It does not interface with the pressurized chemical refrigeration lines, meaning it is not an EPA-regulated task under Section 608 of the Clean Air Act. Anyone can legally replace this part.

Rules for Sourcing a New Replacement Component:

- Match the Microfarad (μF) Rating Exactly: If the original factory label reads 45/5 μF, you must install a replacement rated at exactly 45/5 μF. Never guess or alter microfarad values, as improper sizing will permanently damage the motor windings over time.

- Match or Exceed the Voltage Rating: If your dead capacitor is rated at 370V, you can safely install a replacement rated at 370V or 440V. A higher voltage rating simply means the internal film sheets are thicker and more heavy-duty. However, never downgrade a voltage rating. Installing a 370V capacitor into a system that requires a 440V configuration will cause the component to fail or explode within hours of operation.

- Mount Securely: Ensure the aluminum canister is clamped tightly within its galvanized retaining strap to prevent vibration chafing.

- Land the Terminals Cleanly: Refer directly to the digital photo taken on your phone before disassembly. Land your compressor wire on HERM, your fan wire on FAN, and your primary power feed lines on C.

6.12 What You'll Learn Next

Now that you can test and replace motor run capacitors like an experienced trade professional, you are ready for the next critical diagnostic module: Chapter 7 — Flame Sensors and Ignition Troubleshooting.

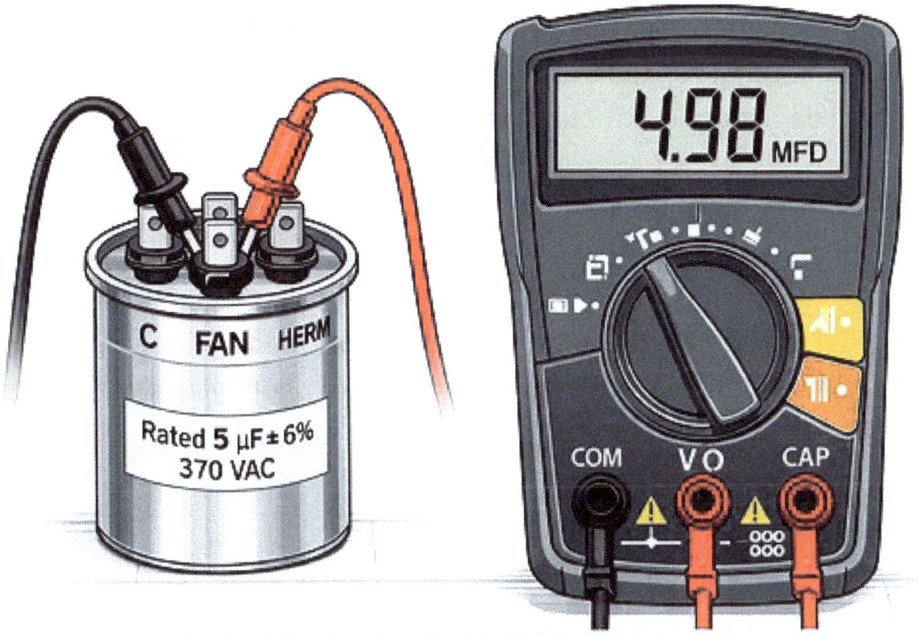

Compare reading to rated value ±6%.

CHAPTER 7 — Switches, Sensors, & Safety Devices

The Hidden Guardians That Keep Climate Systems Running Safely

Every modern heating and air conditioning system is packed with an array of integrated safety devices. Some protect the homeowner from hazardous conditions like fire, toxic gas leaks, or electrical hazards. Others shield expensive equipment—such as the compressor or heat exchanger—from catastrophic structural failure.

Here is the central rule of HVAC diagnostics: when a safety device opens or detects a fault, the main control board halts the ignition or cooling sequence and shuts down the entire system instantly.

This chapter breaks down exactly how these safety switches and sensors operate, how to test them using your digital multimeter, and how to pinpoint the underlying system failures that cause them to trip.

7.1 Limit Switches (The High-Limit)

The Overheat Protector

The high-limit switch is a temperature-sensitive safety device mounted directly on the furnace casing, where its sensing element extends into the supply air plenum or sits adjacent to the heat exchanger cells. Its primary jobs are to shut off the gas burners if the furnace overheats, protect the metallic heat exchanger from cracking due to thermal stress, and eliminate fire hazards.

- Normal State: Normally Closed (NC), allowing a continuous 24V AC safety signal to pass through the circuit to the gas valve.
- When It Trips: If internal plenum temperatures exceed the factory safety threshold, an internal bimetallic disc warps, snapping the electrical contacts open. This immediately cuts power to the gas valve to extinguish the burners, while the indoor blower motor continues running indefinitely to cool down the blistering heat exchanger.

Normal Condition:

24V AC Signal ---> Contacts Closed ---> To Gas Valve

Overheat Condition:

24V AC Signal ---> Contacts Open (Circuit Broken) ---> Power Interrupted

Common Causes of High-Limit Trips

A limit switch rarely trips because the switch itself is broken; it trips because the furnace cannot shed its heat due to poor airflow:

- A heavily impacted, dirty air filter.
- Blocked or furniture-covered return air grilles.
- A dirty blower wheel with packed blades that cannot push air efficiently.
- A weak or failing indoor blower motor running below designed RPM.
- Too many supply registers closed off throughout the house.
- Static pressure issues from poorly designed, undersized ductwork.

How to Test a High-Limit Switch

1. Turn off all power to the furnace.
2. Pull the two wire leads off the limit switch terminals to isolate it from the control circuit.
3. Set your digital multimeter to the Continuity (Beeper) or Resistance (\Omega) setting.
4. Touch your meter leads to the two spade terminals at room temperature.
    - Healthy Reading: The meter should beep cleanly or read 0.0 \Omega (Closed).
    - Failed Reading: If the meter reads OL (Open Loop) while the furnace is completely cold, the internal bimetallic linkage is permanently broken, and the switch must be replaced.
5. Field Airflow Test: If the switch is closed when cold but pops open a few minutes into the heating cycle, leave your meter leads attached and watch the circuit. If it opens mid-cycle, the switch is doing its job—you must troubleshoot the airflow restrictions listed above.

7.2 Rollout Switches

The Flame-Safety Guardians

Rollout switches are thermal safety devices positioned on the exterior perimeter of the burner assembly box. They are designed to detect flame rollout—a dangerous malfunction where the fire fails to draw cleanly into the heat exchanger tunnels and instead spills out into the front mechanical vestibule of the furnace cabinet.

- Normal State: Normally Closed (NC).
- When It Trips: If ambient heat inside the burner cabinet spikes due to escaping flames, a small internal solder link melts or a bimetallic strip pops, opening the contacts. This cuts the 24V AC safety string, drops out the gas valve, and triggers a hard lockout on the primary control board.

CRITICAL SAFETY WARNING: Unlike high-limit switches, rollout switches are almost always manual-reset devices equipped with a small plastic button in the center of their housing. If you find a rollout switch tripped, do not simply push the button to reset it. A tripped rollout switch is a strong indicator of a severe, high-risk mechanical failure. The system must remain off until the root cause is diagnosed and corrected.

Critical Field Diagnostics

If a rollout switch opens, inspect the following system components immediately:

- A cracked or breached heat exchanger cell.
- Internal soot blockages inside the heat exchanger.
- A blocked venting flue or restricted exhaust cap.
- Severely dirty, rusted, or misaligned burners.
- Incorrect manifold gas pressure causing poor draft.

How to Test a Rollout Switch

1. Isolate power and disconnect the wiring leads from the switch.
2. Set your multimeter to Continuity or Low Ohms.
3. Place your test probes across the terminals. A healthy, untripped switch will register a closed circuit (0.0 \Omega). If it registers OL, push the center manual reset button. If it stays open or immediately opens again upon operation, severe combustion backpressure is present.

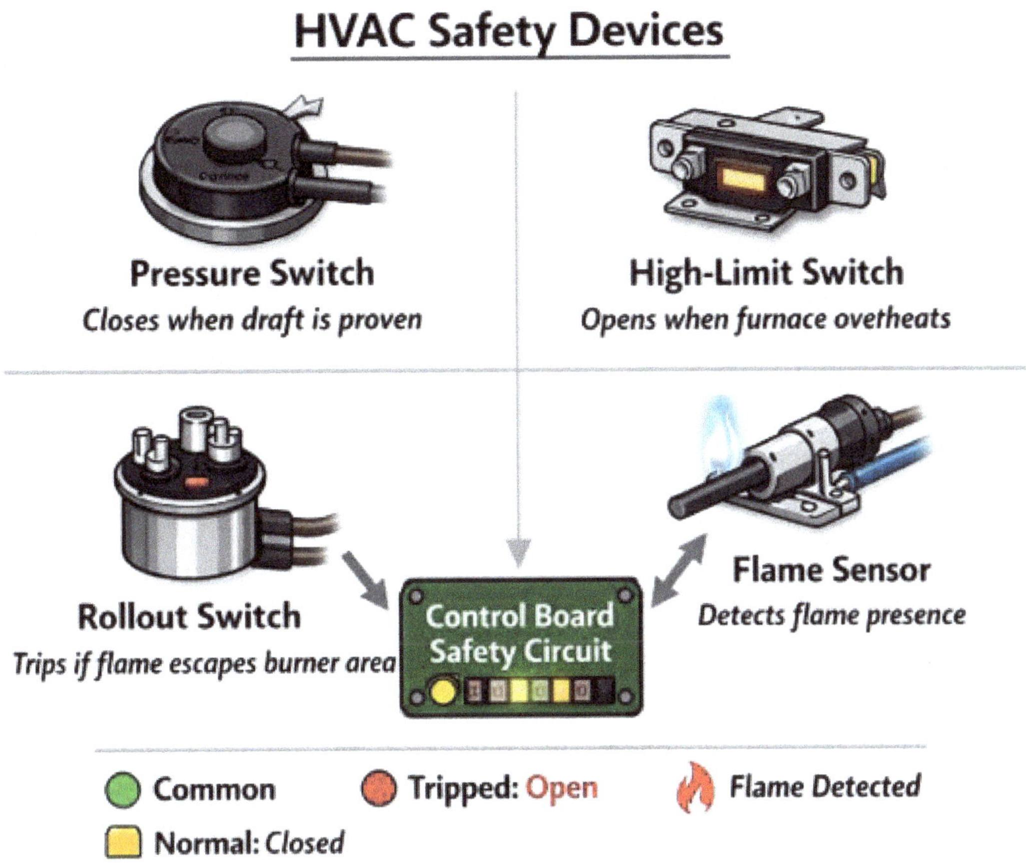

7.3 Pressure Switches

The Inducer Motor's Best Friend

The pressure switch is a pneumatic safety device linked to the draft inducer motor housing via a flexible silicone or rubber sensing tube. It verifies that the inducer fan is spinning fast enough to clear toxic combustion byproducts out of the building venting system before the control board permits ignition.

- Normal State: Normally Open (NO). This is an essential safety check; the control board verifies the switch is open *before* starting the inducer to ensure the contacts aren't welded shut.
- When It Closes: When the draft inducer runs, it creates a localized vacuum (negative static pressure). This suction pulls an internal rubber diaphragm inward, closing the electrical contacts and signaling the board that it is safe to energize the igniter and open the gas valve.

Why a Pressure Switch Fails to Close

If the inducer motor roars to life but the pressure switch fails to close, the problem is rarely a faulty switch. Check these common culprits:

- A clogged condensate drain line or full trap (in 90%+ high-efficiency condensing furnaces).
- A tiny piece of insect debris or rust blocking the small metal port where the hose slips onto the inducer housing.
- A cracked, split, or water-logged rubber pressure tube.
- A bird's nest, ice accumulation, or debris blocking the PVC exhaust or intake pipe outdoors.
- A failing inducer motor that is spinning too slowly to pull the required water column (WC) vacuum pressure.

How to Test a Pressure Switch

Testing a pressure switch requires tracking its state across the furnace's starting sequence:

- Idle Phase (Inducer Off): The switch should be Open, reading OL (Open Loop) on your meter.
- Inducer Startup Phase: The switch transitions, causing your meter reading to bounce from OL to 0.0 \Omega.
- Full Inducer Speed Phase: The switch must be fully Closed, showing a clean, continuous 0.0 \Omega connection.

If your meter reads OL during the entire time the inducer runs, pull the hose, clear the ports with a small wire or paperclip, and verify the vent piping is completely clear before condemning the switch.

7.4 Flame Sensors

The Guardian of Safe Combustion

The flame sensor is a simple, solid stainless steel rod surrounded by a ceramic insulating base, positioned directly in the path of the last burner flame pocket. It acts as an optical-free verification system ensuring gas isn't dumping into the home without burning.

How It Works: Flame Rectification

Flame sensors do not measure temperature. They rely on a physics principle known as flame rectification. Burning gas is highly ionized, meaning it can conduct a tiny amount of electricity. The furnace control board sends an alternating current (AC) voltage out to the flame sensor rod. When the burner ignites, the flame bridges the gap between the small sensor rod and the large, grounded metal chassis of the burner assembly.

Because the ground area is much larger than the sensor rod, the flame acts as a one-way electrical valve, converting the AC voltage into a tiny microamp (\mu\text{A}) Direct Current (DC) signal.

Control Board sends AC Voltage ---> Flame Sensor Rod ---> Passes through the Flame ---> Hits the Grounded Burner Plate ---> Returns to Board as a DC Microamp Signal.

If the control board does not receive a stable DC microamp signal (typically between 1.5 \mu\text{A} and 5.0 \mu\text{A}) within 1 to 3 seconds of the gas valve opening, it shuts off the gas immediately to prevent a toxic vapor buildup, retries a few times, and then locks out.

Symptoms of a Dirty Flame Sensor

- The furnace igniter fires and the burners light perfectly, burn cleanly for exactly 2 to 5 seconds, and then suddenly the gas valve snaps shut.
- The system repeats this brief ignition loop 3 times in a row before going silent and flashing an error code.
- The Cause: Over time, the high-temperature combustion environment coats the steel rod in an invisible layer of non-conductive silica (glass-like carbon buildup). This microscopic insulation layer blocks the path of the microamp current.

How to Service and Clean a Flame Sensor

Cleaning a flame sensor is one of the most common and effective seasonal furnace repairs in the trade:

1. Turn off the electrical power switch at the furnace cabinet wrapper.
2. Remove the single hex-head screw holding the ceramic base to the burner bracket.
3. Gently pull the sensor out of the burner compartment and disconnect the single wire lead.
4. Clean the metal rod using a piece of Scotch-Brite green cleaning pad or fine steel wool. Lightly rub the metal until it shows a bright, clean satin finish.
    - *Note:* Never use rough, heavy-grit sandpaper, as it scratches deep grooves into the steel where carbon can quickly collect, causing premature failure next season.
5. Wipe off any lingering metal dust with a dry cloth, reinstall the assembly, and reconnect the wire.

7.5 Thermistors (Temperature Sensors)

The System's Thermometers

Thermistors are solid-state resistors that alter their internal electrical resistance based on direct exposure to temperature changes. They are most commonly Negative Temperature Coefficient (NTC) sensors, meaning their electrical resistance *drops* as the surrounding temperature *rises*.

HVAC systems deploy these small sensors across various critical operational points:

- Supply & Return Air Sensors: Used in modulating furnaces and heat pumps to calculate real-time temperature splits.
- Evaporator Coil Sensors: Placed between aluminum fins to detect an active coil freeze-up before air blocks completely.
- Liquid/Suction Line Sensors: Clamped to copper refrigerant lines to calculate electronic superheat and subcooling metrics dynamically.

How to Test a Thermistor

To diagnose a erratic or failing thermistor, you must use an Ohmmeter along with the manufacturer's specific Temperature-vs-Resistance Chart:

1. Isolate power and disconnect the thermistor plug from the main control board.
2. Set your multimeter to Ohms (\Omega). Measure the resistance across the two lead wires.
3. Use a separate probe or pocket thermometer to check the actual ambient temperature directly at the sensor head.
4. Cross-reference your readings against the technical manual sheet. If the chart indicates that a temperature of 70°F should register exactly 11,883 \Omega, but your meter reads 2,500 \Omega or OL, the internal semiconductor matrix has failed, and the sensor must be replaced.

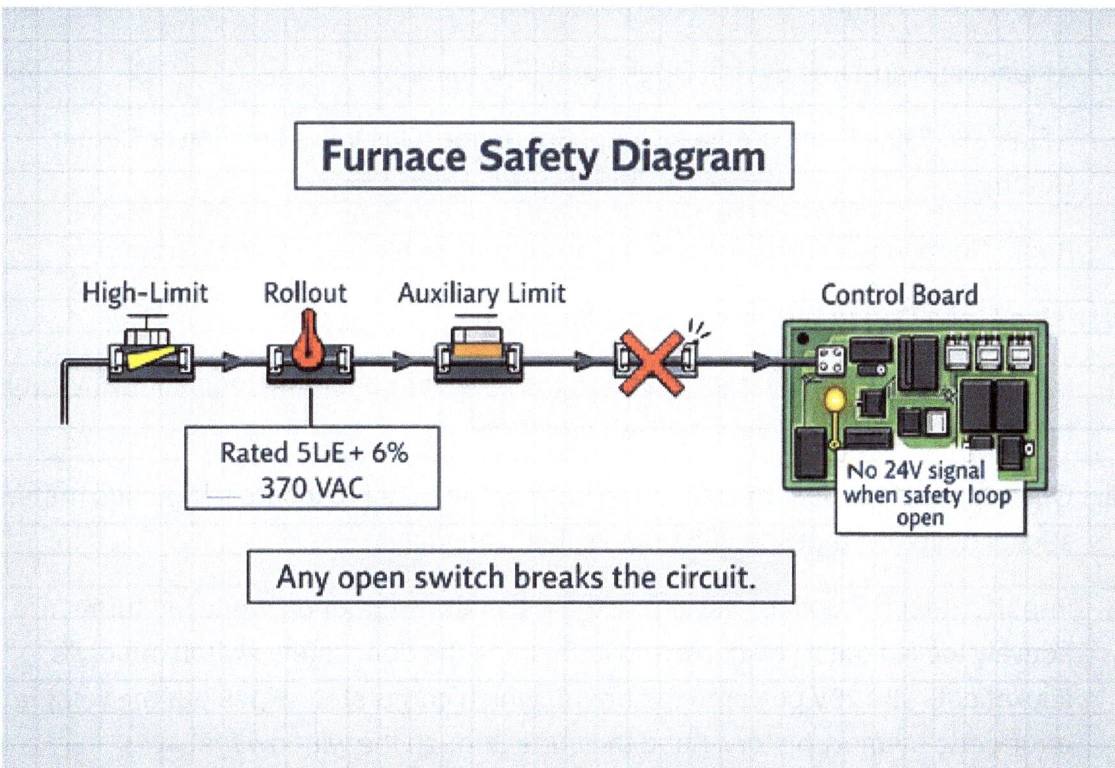

7.6 Control Board Flash Codes

The Furnace's Way of Communicating

When an integrated safety switch breaks the circuit string, the onboard microprocessor stores a specific diagnostic fault code and broadcasts it using a bright red or green LED mounted to the control board surface. This light can be viewed through a small clear plastic inspection window built into the lower blower compartment door.

While exact flash patterns vary between manufacturers (e.g., Carrier, Goodman, Lennox), the underlying diagnostic logic remains universal.

Standard HVAC Diagnostic Codes

- Steady On / Slow Blink: Normal Operation (No Call for Heat)
- 1 Flash: Flame Loss / Ignition Failure
- 2 Flashes: Pressure Switch Stuck Open
- 3 Flashes: Pressure Switch Stuck Closed
- 4 Flashes: High-Limit Switch Open Fault (Overheating)
- 5 Flashes: False Flame Sensed Out of Sequence (Gas Valve Leaking or Circuit Glitch)
- Rapid Continuous Flash: Incorrect Line Polarity or Missing System Ground

How to Use Flash Codes Without Erasing Memory

1. Look through the plastic sight glass on the blower door *before* loosening any cabinet fasteners. Count the repeating flash cadence.
2. Open the cabinet door or look at the back of the panel to locate the plastic matrix sticker decoding that specific brand's flash language.
3. Important Tech Practice: Never pull the main blower door off while the furnace is actively locked out if you can avoid it. Pushing the door safety switch out of its socket cuts all 120V power to the board, which completely wipes out the volatile error-code memory history. Read the code through the window first.

7.7 Safety Devices in Cooling Mode

Protecting the Sealed System

While gas furnaces rely heavily on airflow and combustion safety switches, outdoor air conditioning sections rely on mechanical pressure cutouts installed directly inside the sealed copper refrigerant loops to prevent compressor destruction.

1. High-Pressure Switch (HPS)

The high-pressure cutout is wired in series with the outdoor 24V contactor coil. If head pressure spikes to dangerous thresholds (typically over 450–600 PSI for R-410A systems), the switch opens, breaking power to the contactor and stopping the compressor before it ruptures mechanical seals or burns out its motor insulation.

- *Primary Causes:* A completely blocked, dirt-choked outdoor condenser coil, a dead condenser fan motor, or a severe refrigerant overcharge.

2. Low-Pressure Switch (LPS)

The low-pressure switch protects the compressor from running in a hazardous vacuum or operating with insufficient refrigerant mass flow, which is necessary to cool the compressor motor windings. If suction pressure drops below a safe threshold (typically under 20–50 PSI), the switch opens the contactor circuit.

- *Primary Causes:* A major refrigerant leak, a severely restricted thermal expansion valve (TXV), or a completely frozen indoor evaporator coil caused by an airflow restriction.

7.8 Summary: How Safety Loops Interact

Think of the safety loop as a single, unbroken circle of wire routing out from the 24V AC transformer, traveling through every safety device one after the other, and finally returning back to the control board terminal.

24V Transformer Source ---> High-Limit Switch ---> Rollout Switch ---> Pressure Switch ---> Control Board Safety Input Terminal

If any single switch along this path opens due to heat, flame, or pressure anomalies, the control loop breaks. The main board instantly recognizes this broken path, halts operation, logs the corresponding error code, and keeps the home safe from cascading damage. Understanding these safety interlocks transforms complex troubleshooting into a structured, logical sequence of checking continuity along the chain.

7.9 What You'll Learn Next

Now that you can track safety circuits, test for continuity, and diagnose system faults using control board flash codes, you are ready for the next troubleshooting phase: Chapter 8 — Thermostat Circuitry and Low-Voltage Wiring Control Paths.

CHAPTER 8 — Control Boards & Relays

The Brain of the Furnace — How It Thinks, How It Fails, and How to Diagnose It

If the wall thermostat is the "switchboard" and the low-voltage thermostat wire bundle is the "nervous system," then the integrated furnace control board (IFC) is the brain of the entire HVAC system.

Every single decision the furnace makes—every motor that roars to life, every safety parameter checked, every ignition attempt timed—is calculated and executed by this centralized circuit board.

However, here is a critical industry truth: the vast majority of furnace failures are not caused by a faulty control board. They are caused by something external that the board is correctly reacting to. This chapter teaches you how to map out the board's logical processing, how to execute systematic voltage checks, and how to identify exactly when a control board is truly dead versus when it is simply doing its job by enforcing a safety lockout.

8.1 What the Control Board Actually Does

The integrated control board is a specialized microchip processor engineered to act as a logic engine and power distributor. It manages four primary tasks simultaneously:

1. Gathers System Inputs: It continuously monitors low-voltage command signals from the room thermostat, along with safety status signals from the high-limit switches, rollout sensors, exhaust pressure switches, and flame rectification rods.
2. Executes Logic Decisions: It evaluates the incoming data to answer processing questions: *Has the inducer fan established a safe draft? Is a safety limit open? Is it safe to release highly flammable gas? Has a flame been proven?*
3. Distributes Electrical Power Outputs: It acts as an automated traffic director, routing 24V AC low-voltage current to the gas valve or outdoor compressor contactor, while switching 120V AC line-voltage current to the draft inducer, the hot surface igniter, and the multi-speed blower motor.
4. Disseminates Diagnostic Feedback: It monitors internal timing clocks, logs failed ignition or retry attempts, enforces safety lockouts, and broadcasts diagnostic data via an onboard LED flash indicator.

The board serves as the ultimate referee of the entire sequence of operation.

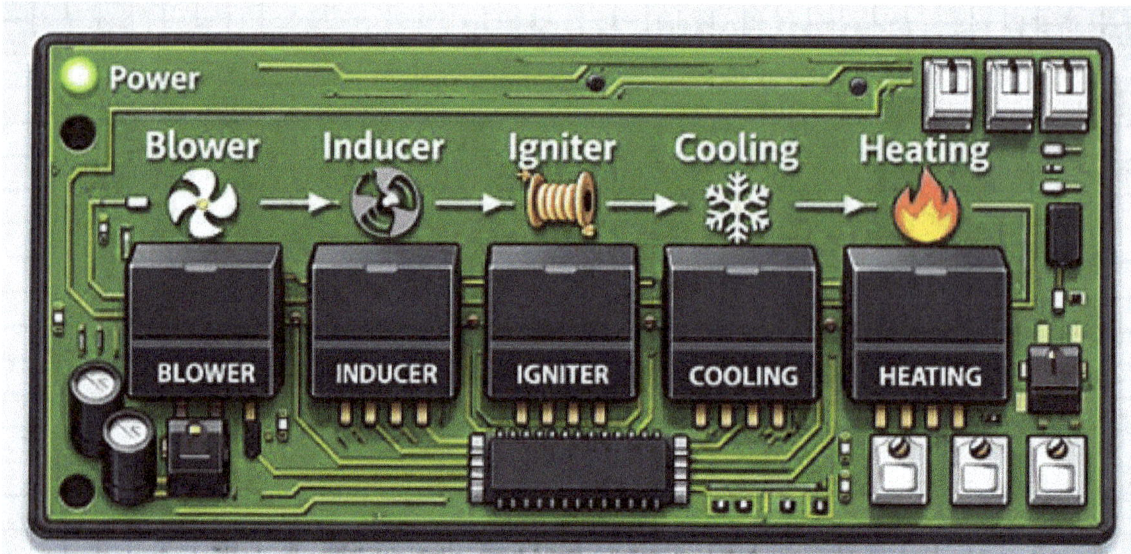

Each relay energizes its respective component when commanded by the control board.

8.2 The Two Sides of a Control Board

An integrated HVAC control board is physically split into two distinct operational territories. Understanding this division is vital for safe and accurate troubleshooting with a multimeter.

A. The Low-Voltage Territory (24V AC)

This side handles system processing, communication, and safety validation.

- Core Components: Thermostat screw terminal strips (R, C, W, Y, G), safety limit string wire harnesses, and the flame sensor terminal.
- Common Control Failures: A blown 3-amp or 5-amp automotive blade fuse, a low-voltage wire shorting flat against a sharp sheet metal casing edge, a dead step-down transformer, or a cracked low-voltage wiring pin connector solder joint.

B. The High-Voltage Territory (120V AC / 240V AC)

This side handles raw mechanical workload distribution.

- Core Components: Main line-voltage power input terminals (Line/L1 and Neutral), spade terminals for motor speed taps (Heat, Cool, Fan, Spare), and the high-voltage step-down transformer primary connection.
- Common Workload Failures: Heavy plastic relays containing contacts that are welded shut or burned open, shattered copper circuit board traces caused by high-amperage surges, or cold solder joints that separate under thermal stress.

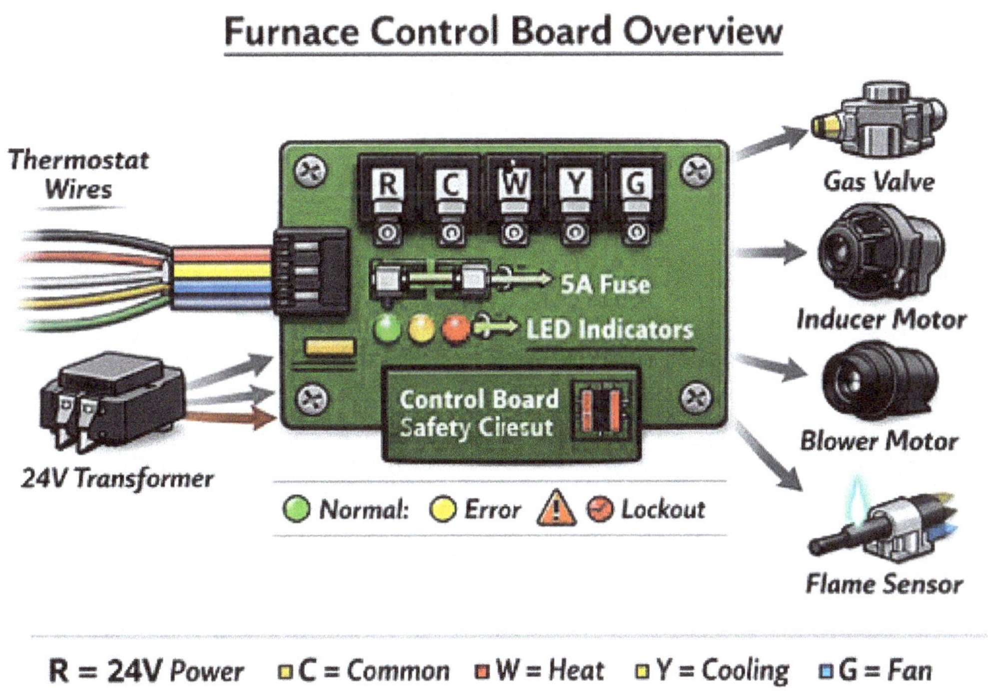

8.3 How Relays Work (The Board's "Muscles")

A relay is simply an electrically operated switch. It allows a low-current, safe signal (like 24V AC from a thermostat) to energize an electromagnetic coil, which pulls a heavy-duty set of mechanical copper contacts closed to route a high-current, dangerous load (like 120V AC or 240V AC) down a wire to a motor.

For example, when the room thermostat calls for heat:

1. The thermostat closes its internal switch, sending 24V AC down the White (W) wire to the control board.
2. The board's microchip reads this input and routes 24V AC directly to the internal draft inducer relay coil.
3. The miniature coil turns into a magnet, snapping the heavy 120V contacts inside the plastic relay cube closed.
4. Line-voltage electricity (120V AC) flows out of the terminal straight to the inducer motor, causing it to spin.

Integrated boards utilize dedicated onboard relays to cycle the combustion inducer fan, the hot surface igniter element, the indoor circulation blower, and the auxiliary electric heat strips found in heat pump air handlers.

Symptoms of a Failed Relay

Because relays are physical mechanical switches, they eventually wear out after thousands of duty cycles:

- Contacts Burned Open: The control board clicks audibly when a component should start, but your multimeter reads 0V at the output terminal. The motor or igniter never receives power.
- Contacts Welded Closed: A motor runs continuously without stopping. For instance, the draft inducer fan or main blower motor runs the instant the main breaker is flipped on, even with the thermostat completely disconnected from the wall.
- Coil Burnout: The microchip attempts to energize the relay, but the relay fails to make any physical clicking sound, and no power passes across the switch points.

8.4 How to Test a Control Board (The Professional Method)

You do not test a control board by probing random spots on its circuit board traces. Instead, you test the board by verifying its inputs and measuring its outputs using a logical sequence aligned with the furnace's sequence of operation.

Ensure your digital multimeter is set to AC Voltage. Keep your black probe anchored to a verified system ground or the white Neutral terminal block when checking high-voltage lines, or to the C (Common) terminal when checking low-voltage lines.

Test 1 — Confirm Main Supply Power (120V AC Input)

Measure voltage across the incoming main line terminal (L1) and the Neutral block.

- The Target: You must read a steady 110V AC to 125V AC.
- If 0V: The fault is upstream. Check for a tripped circuit breaker, an off service switch, a loose wire nut in the furnace junction box, or a wide-open blower door safety interlock switch.

Test 2 — Confirm Control Loop Power (24V AC Input)

Measure voltage directly across the screw terminals labeled R and C on the thermostat strip.

- The Target: You must read 24V AC to 28V AC.
- If 0V: Check the small automotive blade fuse pressed into the face of the board. If the fuse is blown, look for a low-voltage wire shorting to ground before replacing it. If the fuse is intact but R-to-C reads 0V, your step-down transformer has failed.

Test 3 — Check the Combustion Inducer Output (120V AC Output)

Set the thermostat to call for heat. The moment the thermostat clicks, measure across the active INDUCER or IND spade terminal on the board and the Neutral block.

- The Target: You should read 120V AC.
- The Diagnosis: If your meter reads 120V AC but the inducer motor does not spin, the board is healthy; the inducer motor winding is burned out or seized. If your meter reads 0V AC but the board is flashing a "Call for Heat" light or clicking, the board's internal inducer relay is broken.

Test 4 — Check the Igniter Output (120V AC Output)

Wait for the inducer to ramp up and the pressure switch circuit to close. At the exact moment in the sequence where ignition should occur, measure voltage across the two pins of the igniter wire harness block coming off the board.

- The Target: You should read 120V AC (or the manufacturer's specified voltage, such as 80V AC on certain specialized silicon nitride circuits).
- The Diagnosis: If you read full voltage but the igniter stays dark, the board is delivering power perfectly; the igniter element is cracked and dead. If your meter reads 0V AC during the designated ignition window, the board's igniter circuit has failed.

Test 5 — Check the Gas Valve Power (24V AC Output)

The moment the igniter reaches its peak temperature, measure across the GV terminal on the board and the chassis ground or C screw.

- The Target: You must read 24V AC for a brief 4-to-7 second ignition window.
- The Diagnosis: If you read 24V AC but no fire appears, the board is healthy; the gas valve solenoid is bad, or the manual gas valve knob is turned off. If you read 0V AC, the board is intentionally blocking ignition because an upstream safety device (like a pressure or limit switch) opened and broke the chain.

Test 6 — Check the Main Circulation Blower (120V AC Output)

Following successful burner ignition, wait for the board's fan-on delay timer to expire (usually 30 to 45 seconds). Measure between the HEAT speed tap terminal on the board and the Neutral block.

- The Target: You must read 120V AC.
- The Diagnosis: If you read 120V AC but the blower fan sits dead, the board is healthy; the blower motor or its run capacitor has failed. If you read 0V AC, the internal fan relay on the control board is defective.

8.5 When the Control Board Is NOT the Problem

Homeowners and inexperienced technicians frequently condemn and replace functional control boards out of frustration. A control board is operating exactly as designed and is not defective if it locks out the system due to any of the following field conditions:

- The draft inducer runs, but a pressure switch refuses to close due to a blocked PVC exhaust vent pipe or a clogged condensate trap.
- The burners light but extinguish after 3 seconds because a dirty, oxidized flame sensor rod cannot return a microamp current back to ground.
- The system shuts down mid-cycle because a clogged, dust-choked air filter caused the heat exchanger to trip the high-limit switch.
- The outdoor cooling unit fails to turn on because a low-pressure safety switch opened to protect a compressor that leaked all its refrigerant.

The control board is shutting down the equipment on purpose to keep the building safe. Fix the underlying mechanical or airflow issue first.

8.6 When the Control Board IS the Problem

An integrated control board is genuinely defective and must be replaced if it exhibits any of the following symptoms:

1. Correct Inputs with Zero Output: You verify 24V AC across R and W, all safety limit switches are proven closed with your ohmmeter, and 120V input power is verified, yet the board sits completely inert and refuses to route power to the inducer motor or log an error code.
2. Stuck or Shattered Relays: An internal motor runs non-stop without a command signal, or a relay clicks loudly but fails to pass line voltage across its internal contacts.
3. Obvious Physical Degradation: Visual inspection reveals charred black scorch marks on the fiberglass surface, melted plastic relay frames, broken copper circuit board tracks, or leaking, swollen capacitors.
4. Erratic or Random System Behavior: The furnace loops through random operational phases without any input from the thermostat—such as cycling the igniter on and off out of sequence or running the main blower motor continuously with no call for heat or fan present.
5. A Completely Dead Diagnostic LED: Main supply voltage is present, the low-voltage transformer is verified to be outputting a clean 24V AC signal to the board, but the onboard diagnostic LED remains dark and refuses to display a status or heartbeat flash.

8.7 How to Replace an Integrated Control Board Safely

Replacing an internal furnace or air handler control board is an electrical component swap. It does not interface with gas line connections or pressurized chemical refrigerant loops, meaning it is not an EPA-regulated task. Anyone can legally replace this part.

Execution Steps:

1. Isolate All Electrical Power: Flip the double-pole AC breaker to the OFF position and verify zero voltage is present inside the furnace cabinet with your multimeter.
2. Document the Original Layout: Take a high-resolution photograph of the old board with your phone. This photo serves as your primary reference guide for wire positioning.

3. Label Critical Wiring Bundles: If necessary, apply small strips of tape to identify matching wire runs, especially individual motor speed taps (Heat, Cool, Fan) and low-voltage thermostat groupings.
4. Extract the Old Board: Disengage the plastic snap-standoff pins or remove the grounding screws holding the board frame to the metal blower housing shelf.
5. Install the New Replacement Board: Snap the new board securely onto the mounting standoffs. Ensure it sits perfectly firm and does not touch any surrounding sheet metal.
6. Transfer Connections Systematically: Move wires over from the old board to the new board one single wire at a time. This technique eliminates the risk of miswiring complex multi-pin connectors.
7. Verify Voltage Compatibility: Ensure the small onboard jumper configurations or delay-setting dip switches match the airflow parameters outlined in your original equipment manual.
8. Re-energize and Test: Reinstall the blower compartment door to depress the safety switch, restore electrical power, and run the system through a full heating and cooling sequence to verify flawless operation.

8.8 What You'll Learn Next

Now that you understand the logic brain of the heating plant, you are ready to move into: Chapter 9 — Gas Valves, Manifold Pressures, and Safe Combustion Testing Loops.

CHAPTER 9 — Airflow: The Silent Killer of HVAC Systems

Why 80% of HVAC Problems Aren't Electrical — They're Airflow

If you ask any seasoned HVAC technician what destroys more furnaces and air conditioners than anything else, they'll tell you the same thing:

Airflow — or the lack of it — is the silent killer of HVAC systems.

Electrical problems are dramatic.
Capacitors explode.
Contactors burn.
Transformers blow fuses.

But airflow problems?
They creep in slowly.
Quietly.
And they ruin equipment long before homeowners realize something is wrong.

This chapter teaches you how airflow works, how it fails, and how to diagnose airflow issues with nothing more than your eyes, your hands, and your ears.

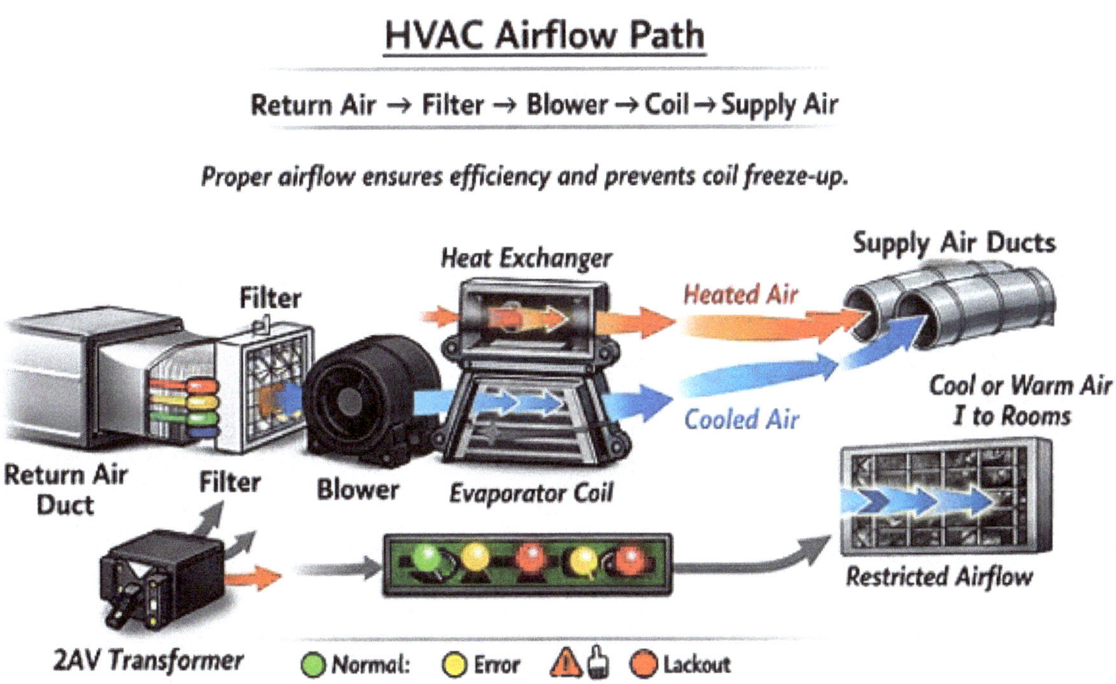

9.1 Why Airflow Matters More Than Most People Realize

Every HVAC system depends on proper airflow to:

- Move heat
- Prevent overheating
- Prevent freezing
- Maintain efficiency
- Protect motors
- Protect the heat exchanger
- Prevent compressor damage

When airflow drops, everything else suffers.

In heating mode:

Low airflow → heat exchanger overheats → limit switch trips → furnace shuts down.

In cooling mode:

Low airflow → evaporator coil gets too cold → coil freezes → compressor overheats.

Airflow is the foundation of HVAC performance.

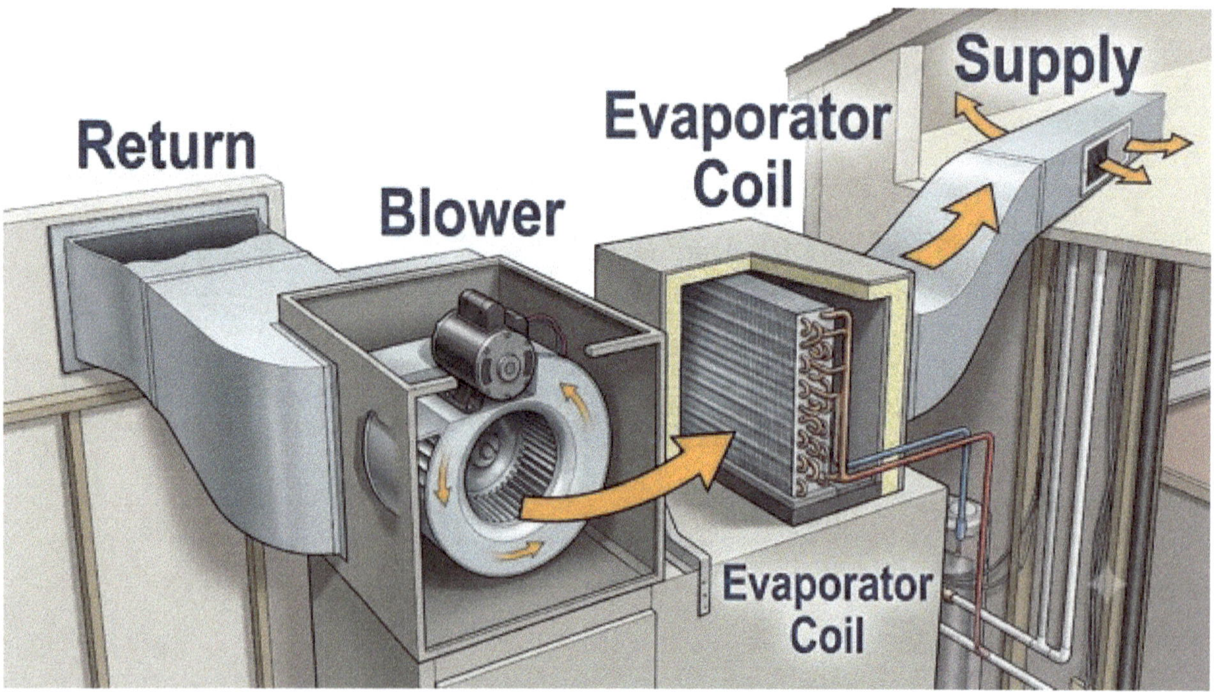

9.2 The Airflow Path (What Air Actually Does)

Airflow in a typical system follows this path:

1. **Return vents** pull air from the home
2. **Return ductwork** carries air to the furnace
3. **Filter** removes dust
4. **Blower motor** pushes air through
5. **Evaporator coil** (cooling) or **heat exchanger** (heating) conditions the air
6. **Supply ductwork** distributes air back into the home
7. **Supply vents** deliver conditioned air

If any part of this path is restricted, the entire system suffers.

9.3 The #1 Airflow Killer: Dirty Air Filters

A dirty filter is the most common cause of:

- Furnace overheating
- AC freeze-ups
- High energy bills
- Weak airflow
- Short cycling
- Blower motor strain

How often should filters be changed?

- 1" filters: every 30–60 days
- 2" filters: every 60–90 days
- 4–5" media filters: every 6–12 months

Warning:

High-MERV filters can choke airflow if the system wasn't designed for them.

9.4 The #2 Airflow Killer: Blocked Return Vents

Return vents are just as important as supply vents.

Blocked returns cause:

- Furnace overheating
- High-limit trips
- Weak airflow
- Excessive blower noise
- Reduced efficiency

Common causes:

- Furniture pushed against returns
- Rugs covering floor returns
- Homeowners closing too many supply vents

A system must breathe freely to work properly.

9.5 The #3 Airflow Killer: Dirty Evaporator Coils

The evaporator coil sits above the furnace and removes heat and humidity.

Over time, it collects:

- Dust
- Pet hair
- Cooking oils
- Mold
- Construction debris

Symptoms of a dirty evaporator coil:

- AC runs constantly
- Weak airflow
- Coil freezing
- Water leaking from furnace
- High energy bills

A dirty coil can reduce airflow by **30–50%**.

9.6 The #4 Airflow Killer: Dirty Blower Wheels

The blower wheel is the "fan" that moves air through the system.

When it gets dirty:

- Airflow drops
- Motor amps increase
- Furnace overheats
- AC freezes

Symptoms:

- Loud blower
- Weak airflow
- Dust buildup on vents
- High-limit trips

A blower wheel packed with dust can lose **40% of its airflow capacity**.

9.7 The #5 Airflow Killer: Undersized or Poorly Designed Ductwork

Many homes have ductwork that was:

- Installed cheaply
- Designed poorly
- Never sized correctly
- Modified by previous homeowners

Symptoms of bad ductwork:

- Hot and cold rooms
- Loud airflow
- Weak airflow
- Furnace short cycling
- AC freezing

Ductwork is the most overlooked part of HVAC systems.

9.8 How to Diagnose Airflow Problems (Without Tools)

You don't need gauges or fancy equipment to diagnose airflow issues.

Use your hands:

Feel the airflow at supply vents.

Use your ears:

Listen for whistling, roaring, or rattling.

Use your eyes:

Check filters, blower wheels, and coils.

Use your brain:

Compare airflow between rooms.

Use your senses:

If something feels off, it usually is.

9.9 How to Diagnose Airflow Problems (With Simple Tools)

1. Temperature Rise Test (Heating Mode)

Measure supply air vs. return air.

Typical rise: **40–70°F**
If rise is too high → airflow problem.

2. Temperature Drop Test (Cooling Mode)

Measure supply air vs. return air.

Typical drop: **16–22°F**
If drop is too low → airflow problem.
If drop is too high → coil freezing.

3. Visual Inspection

- Filter
- Blower wheel
- Evaporator coil
- Return vents
- Supply vents

9.10 The Silent Killer: Static Pressure

Static pressure is the resistance to airflow inside the duct system.

High static pressure causes:

- Blower motor strain
- Heat exchanger cracks
- AC freeze-ups
- High energy bills
- Loud airflow
- Premature equipment failure

You don't need gauges to suspect high static pressure — the symptoms are obvious.

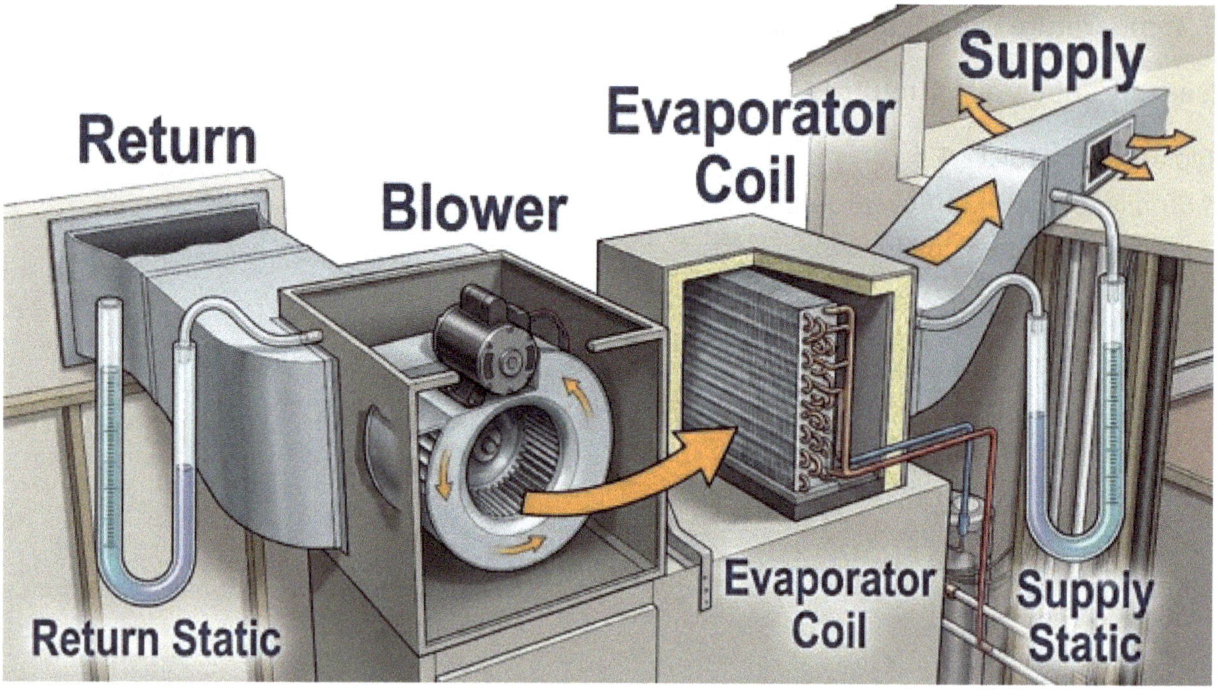

9.11 The Technician's Airflow Checklist

Whenever diagnosing airflow, check these in order:

1. Filter
2. Return vents
3. Supply vents
4. Blower wheel
5. Evaporator coil
6. Ductwork
7. Blower motor speed
8. Heat exchanger cleanliness

This checklist solves 80% of airflow problems.

9.12 What You'll Learn Next

Now that you understand airflow — the most overlooked part of HVAC — you're ready for hands-on maintenance.

Next chapter:

CHAPTER 10 — Deep Cleaning the Outdoor Condenser

The Most Important AC Maintenance Task You Can Do Without Touching Refrigerant

If you only performed **one** maintenance task on your air conditioner every year, it should be this one.

A dirty outdoor condenser coil is responsible for:

- Weak cooling
- High energy bills
- Compressor overheating
- Tripped breakers
- Shortened equipment lifespan
- Loud operation
- Constant running in hot weather

And here's the good news:

Cleaning a condenser is safe, legal, and requires no EPA certification — as long as you don't open the refrigerant system.

This chapter teaches you the exact method technicians use to clean a condenser coil without damaging fins, motors, or electrical components.

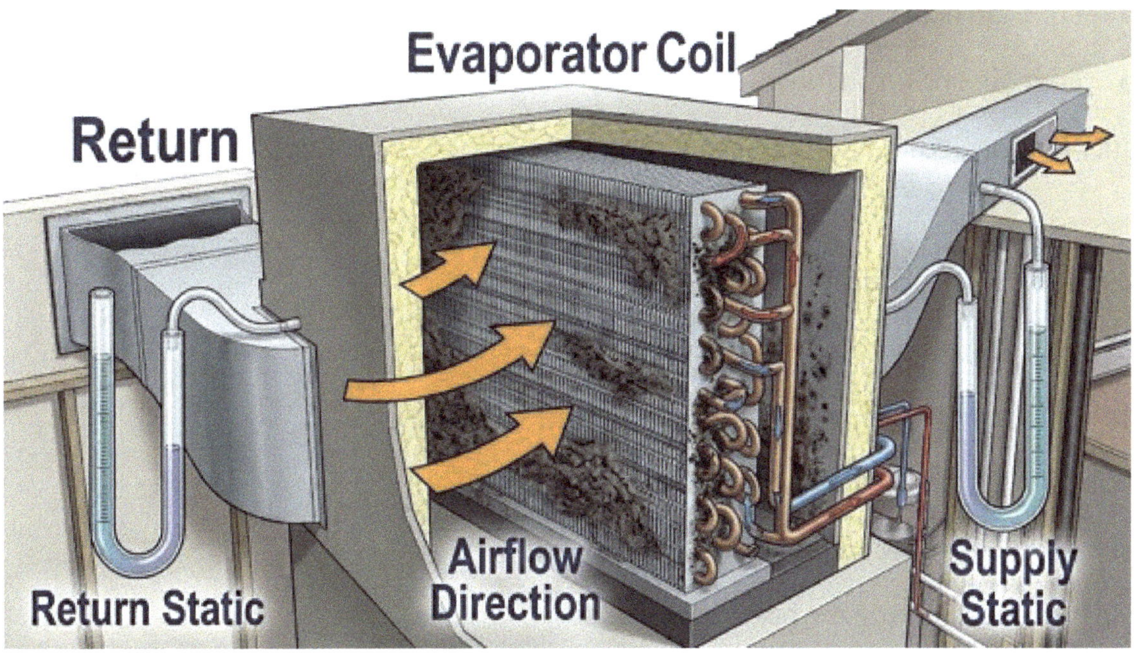

10.1 What the Outdoor Condenser Actually Does

The outdoor unit has one job:

Reject heat from inside the home to the outside air.

It does this by:

1. Pumping hot refrigerant into the condenser coil
2. Using the fan to pull outdoor air through the coil
3. Releasing heat into the environment

If the coil is dirty, the system can't get rid of heat — and everything suffers.

10.2 Symptoms of a Dirty Condenser Coil

A dirty coil causes:

- AC running constantly
- Warm air from vents
- High energy bills
- Compressor overheating
- Outdoor unit extremely hot to the touch
- Loud fan noise
- Poor cooling on hot days

If your AC struggles in July, this is usually why.

10.3 Safety First — Isolating 240V Power

Before touching the outdoor unit, you must kill power.

Step 1 — Locate the Disconnect Box

Mounted on the wall near the condenser.

Step 2 — Pull the Disconnect Block

Or flip the breaker inside the box.

Step 3 — Verify Power Is Off

Use your multimeter to check for 240V at the contactor.

Never trust the disconnect alone — always verify.

10.4 Tools & Materials Needed

- Garden hose with spray nozzle
- Fin comb (optional)
- Coil cleaner (non-acidic)
- Nut driver or drill
- Soft brush
- Gloves
- Safety glasses

No pressure washers — they destroy fins.

10.5 How to Remove the Top Fan Assembly

Most condensers require removing the top panel to access the coil interior.

Steps:

1. Remove screws around the top panel
2. Lift the fan assembly straight up
3. Support it without stressing the wires
4. Do **not** disconnect the fan unless necessary

The fan wires are fragile — treat them gently.

10.6 How to Clean the Condenser Coil (Technician Method)

Step 1 — Remove Debris

- Leaves
- Grass clippings
- Dirt
- Spider webs

Use your hands or a soft brush.

Step 2 — Apply Coil Cleaner

Use a **non-acidic** cleaner designed for residential units.

- Spray from the inside outward
- Let it foam for 5–10 minutes
- Do not let it dry completely

This loosens dirt deep inside the fins.

Step 3 — Rinse From the Inside Out

This is the most important step.

- Spray water from the **inside** of the coil outward
- Use gentle pressure
- Rinse until water runs clear

Spraying from the outside pushes dirt deeper into the coil — always rinse from the inside.

Step 4 — Straighten Bent Fins (Optional)

Use a fin comb to straighten any crushed areas.

Bent fins reduce airflow and efficiency.

Step 5 — Reassemble the Unit

- Lower the fan assembly carefully
- Reinstall screws
- Ensure no wires are pinched

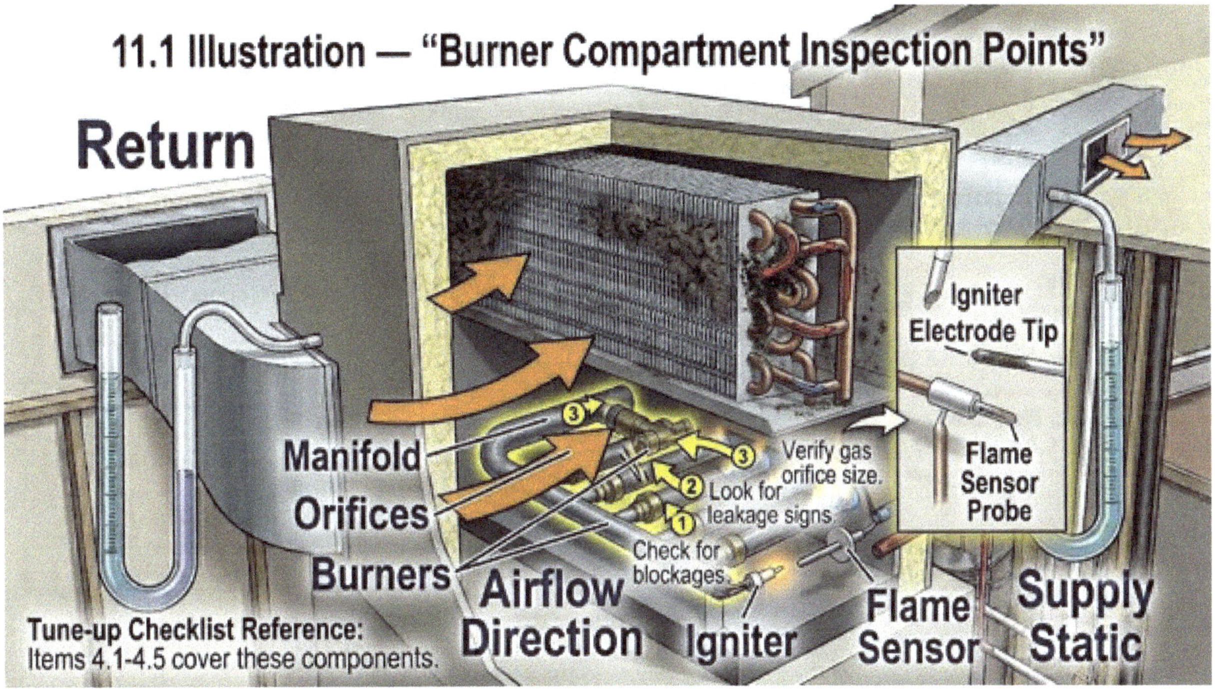

Step 6 — Restore Power

Insert the disconnect block or flip the breaker.

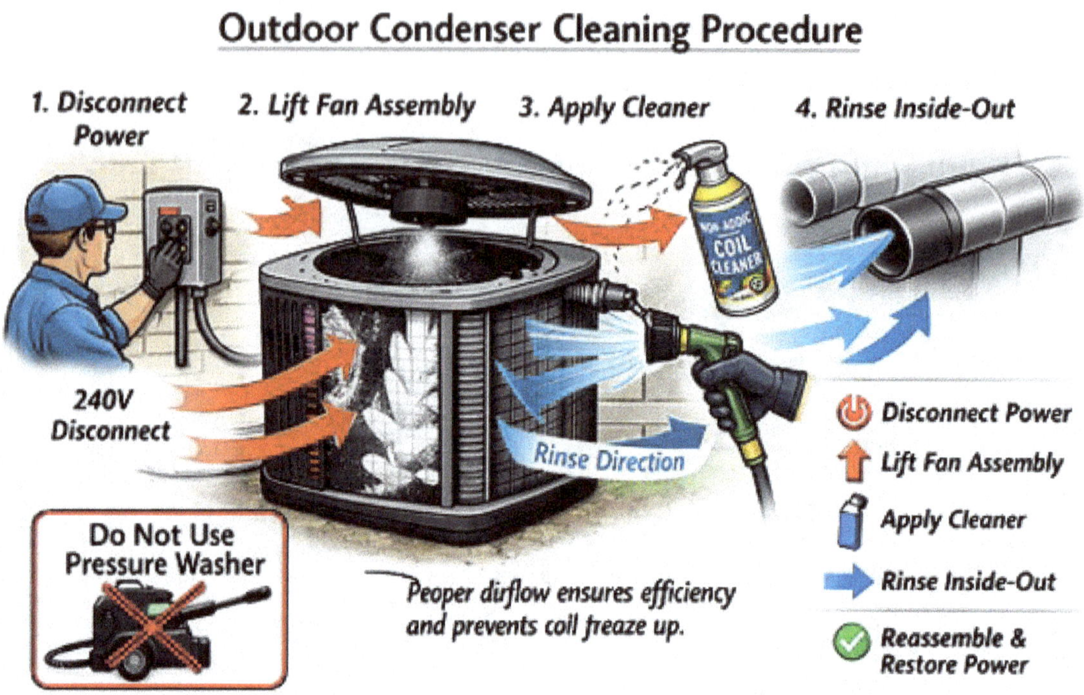

10.7 How Often Should You Clean the Condenser?

- **Once per year** for most homes
- **Twice per year** if you have:
    - Cottonwood trees
    - Pets
    - Construction nearby
    - Heavy pollen

A clean condenser can reduce energy use by **10–20%**.

10.8 What NOT to Do

Never use a pressure washer

It will flatten fins instantly.

Never bend the fan blades

This causes vibration and motor failure.

Never spray electrical components

Control boards and capacitors hate water.

Never open refrigerant lines

EPA-regulated — not covered in this book.

10.9 The Technician's Shortcut: The Hand Test

After cleaning:

- Place your hand above the condenser fan
- Air should be **hot** — very hot
- If air is only warm → coil still dirty or compressor weak

This simple test tells you a lot.

10.10 What You'll Learn Next

Now that you know how to maintain the outdoor unit, it's time to move indoors.

Next chapter:

CHAPTER 11 — Furnace Tune-Up Protocols

The Step-By-Step Maintenance Routine That Keeps a Furnace Running Safely and Efficiently

A furnace tune-up is one of the most valuable maintenance tasks a homeowner can perform — and one of the most misunderstood.
Most people think a tune-up is about "cleaning the furnace," but technicians know the truth:

A proper tune-up is a safety inspection, airflow verification, ignition check, and cleaning protocol all in one.

This chapter gives you the exact sequence technicians use during a professional tune-up.
No fluff.
No unnecessary steps.
Just the real-world process that keeps a furnace running safely and efficiently.

11.1 Safety First — Before You Touch Anything

1. Kill power at the furnace switch

This is usually a light-switch-style disconnect mounted on the side of the furnace.

2. Remove the furnace doors

Most furnaces have an upper (burner) and lower (blower) door.

3. Verify power is off

Use your meter to confirm **0 volts** at the control board.

4. Shut off gas only if needed

Most tune-ups do not require shutting off gas.

5. Never bypass safety switches

Not even for "just a second."

11.2 Step 1 — Inspect and Replace the Air Filter

The filter is the #1 cause of furnace overheating.

What to look for:

- Dirt buildup
- Collapsed filter
- Wrong size
- High-MERV filter choking airflow

Replace if:

- You can't see light through it
- It's been more than 60–90 days
- It's visibly clogged

A clean filter protects the heat exchanger and blower motor.

11.3 Step 2 — Clean the Flame Sensor

A dirty flame sensor is the #1 cause of furnaces that light and then shut off.

How to clean it:

1. Remove the flame sensor (usually one screw)
2. Gently clean the rod with a **Scotch-Brite pad**
3. Do NOT use sandpaper — it scratches the metal
4. Reinstall and tighten screw

This simple step prevents countless no-heat calls.

11.4 Step 3 — Inspect the Hot Surface Igniter (HSI)

The igniter is fragile and cracks easily.

What to look for:

- White spots
- Cracks
- Burn marks
- Loose plug connection

Do NOT touch the igniter with bare fingers

Oil from skin shortens its life.

If it looks questionable, replace it — igniters fail suddenly.

11.5 Step 4 — Check the Inducer Motor and Pressure Ports

The inducer must create proper draft for safe ignition.

What to inspect:

- Inducer wheel for debris
- Pressure switch tubing for cracks
- Inducer port for blockage
- Water in tubing (condensing furnaces)

Clean the inducer port

Use a small drill bit or paper clip to clear the tiny hole.

A clogged inducer port is a top cause of ignition failure.

11.6 Step 5 — Inspect Burners and Burner Crossover

Burners must ignite smoothly and evenly.

What to look for:

- Rust
- Soot
- Debris
- Blocked crossover channels

Cleaning method:

- Remove burners
- Brush lightly with a wire brush
- Vacuum debris
- Reinstall carefully

Smooth ignition prevents delayed ignition "booms."

11.7 Step 6 — Inspect the Heat Exchanger (Visual Only)

You are NOT opening the heat exchanger — that requires advanced tools.

What you CAN do:

- Use a flashlight
- Look for rust
- Look for soot streaks
- Look for hot spots
- Look for discoloration
- Look for cracks (rare but serious)

If anything looks suspicious, call a licensed technician.

11.8 Step 7 — Inspect the Blower Motor and Wheel

The blower is responsible for all airflow.

What to check:

- Dust buildup on wheel
- Loose set screw
- Wobbling wheel
- Motor overheating
- Motor noise
- Capacitor condition

Cleaning the blower wheel:

- Remove blower assembly
- Brush and vacuum
- Avoid bending blades

A dirty blower wheel reduces airflow by up to **40%**.

11.9 Step 8 — Inspect the Condensate Drain System (High-Efficiency Furnaces)

Condensing furnaces produce water.

Check:

- Drain trap
- Drain tubing
- Condensate pump (if present)
- Blockages
- Algae buildup

Clear the drain:

- Flush with warm water
- Add a small amount of vinegar
- Ensure trap is filled after cleaning

A clogged drain can shut down the furnace.

11.10 Step 9 — Check Electrical Connections

Loose connections cause:

- Overheating
- Intermittent shutdowns
- Blower failure
- Ignition failure

Check:

- Control board terminals
- Ground connections
- Wire nuts
- Motor plugs
- Transformer connections

Tighten gently — don't overtighten.

11.11 Step 10 — Perform a Temperature Rise Test

This verifies airflow and heat exchanger performance.

How to test:

1. Run furnace for 10 minutes
2. Measure return air temperature
3. Measure supply air temperature
4. Subtract the two

Typical rise:

40–70°F (check furnace label)

If rise is too high:

- Airflow problem
- Dirty filter
- Dirty blower wheel
- Dirty coil
- Closed vents

If rise is too low:

- Blower speed too high
- Gas pressure low (requires licensed tech)

This is one of the most important tune-up tests.

11.12 Step 11 — Verify Safe Ignition Sequence

Watch the entire sequence:

1. Thermostat calls for heat
2. Inducer starts
3. Pressure switch closes
4. Igniter glows
5. Gas valve opens
6. Burners ignite
7. Flame sensor verifies
8. Blower starts

If any step fails, refer back to Chapters 2 and 7.

11.13 Step 12 — Final Safety Check

Before closing the furnace:

- Ensure all wires are secure
- Ensure all tubing is connected
- Ensure no tools are left inside
- Ensure blower door switch is aligned
- Ensure both doors are installed correctly

A loose blower door will shut down the furnace instantly.

11.14 What You'll Learn Next

Now that you've mastered furnace tune-ups, it's time to move to the cooling side of indoor maintenance.

Next chapter:

CHAPTER 12 — AC Evaporator Coil & Condensate Maintenance

The Hidden Side of AC Maintenance — Quiet, Critical, and Often Ignored

If the outdoor condenser is the "muscle" of your AC system, the **evaporator coil** is the "heart."
It's where the magic happens — where heat is absorbed from your home and transferred to the refrigerant.

But here's the problem:

The evaporator coil is the most neglected component in residential HVAC systems.

It's hidden.
It's hard to access.
And when it gets dirty, the entire system suffers.

This chapter teaches you how to safely access, inspect, and maintain the evaporator coil and condensate system — without touching refrigerant or performing EPA-restricted tasks.

12.1 What the Evaporator Coil Actually Does

The evaporator coil sits above (or beside) the furnace in the indoor air handler.

Its job:

- Absorb heat from indoor air
- Dehumidify the home
- Transfer heat to the refrigerant
- Deliver cool, dry air back into the ducts

When airflow is good and the coil is clean, the system runs efficiently.

When the coil is dirty, everything falls apart.

12.2 Symptoms of a Dirty or Blocked Evaporator Coil

A dirty coil causes:

- Weak airflow
- AC running constantly
- High energy bills
- Coil freezing
- Water leaking from furnace
- Musty smells
- Hot and cold spots in the home

If your AC freezes up, this is the first place to look.

12.3 Safety First — Before Accessing the Coil

1. Turn off power at the furnace switch

This kills the blower and control board.

2. Turn off power at the outdoor disconnect

Prevents the AC from starting unexpectedly.

3. Verify power is off

Use your meter — never assume.

4. Do NOT open refrigerant lines

EPA-regulated.
Not covered in this book.

12.4 How to Access the Evaporator Coil

Most coils are inside a sheet-metal housing called the **A-coil cabinet**.

Steps:

1. Remove screws from the coil access panel
2. Carefully lift or slide the panel off
3. Avoid bending refrigerant lines
4. Do not pry or force anything

Some coils are sealed with foil tape — remove tape gently and reapply later.

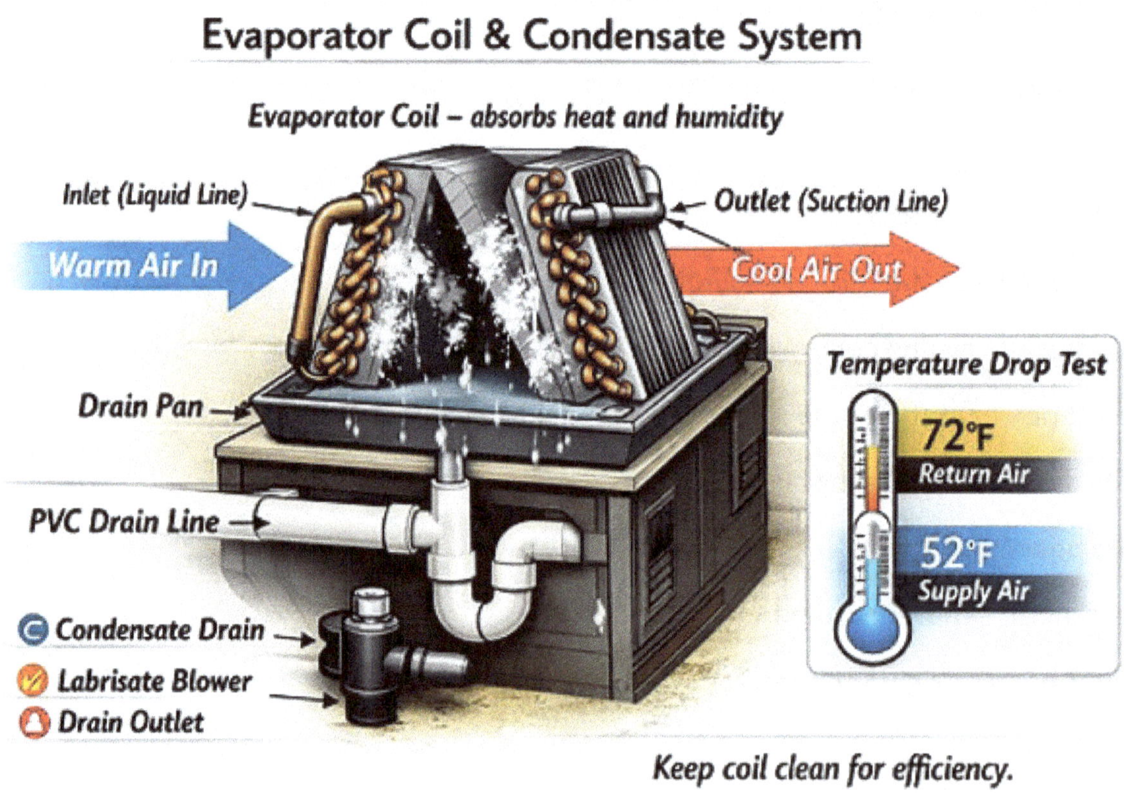

12.5 What a Healthy Coil Looks Like

A clean coil should have:

- Shiny aluminum fins
- No dust buildup
- No mold
- No ice
- No rust streaks
- No oil stains (oil = refrigerant leak → call a licensed tech)

If the coil looks like a lint trap, it's overdue for cleaning.

12.6 How to Clean the Evaporator Coil (Surface Cleaning Only)

You are NOT removing the coil.
You are NOT opening refrigerant lines.

Tools Needed:

- Soft brush
- Vacuum with brush attachment
- Coil-safe cleaner (non-acidic, indoor-rated)
- Spray bottle
- Rags

Steps:

1. Gently brush loose dust from the coil
2. Vacuum debris without crushing fins
3. Spray coil cleaner lightly
4. Let it sit for 5–10 minutes
5. Rinse with a small amount of water (if allowed by manufacturer)
6. Wipe excess moisture

Do NOT:

- Use high-pressure water
- Use acidic cleaners
- Bend fins
- Spray electrical components

This is a light cleaning — enough to restore airflow without risking damage.

12.1 Illustration — "Evaporator Coil Airflow Orientation"

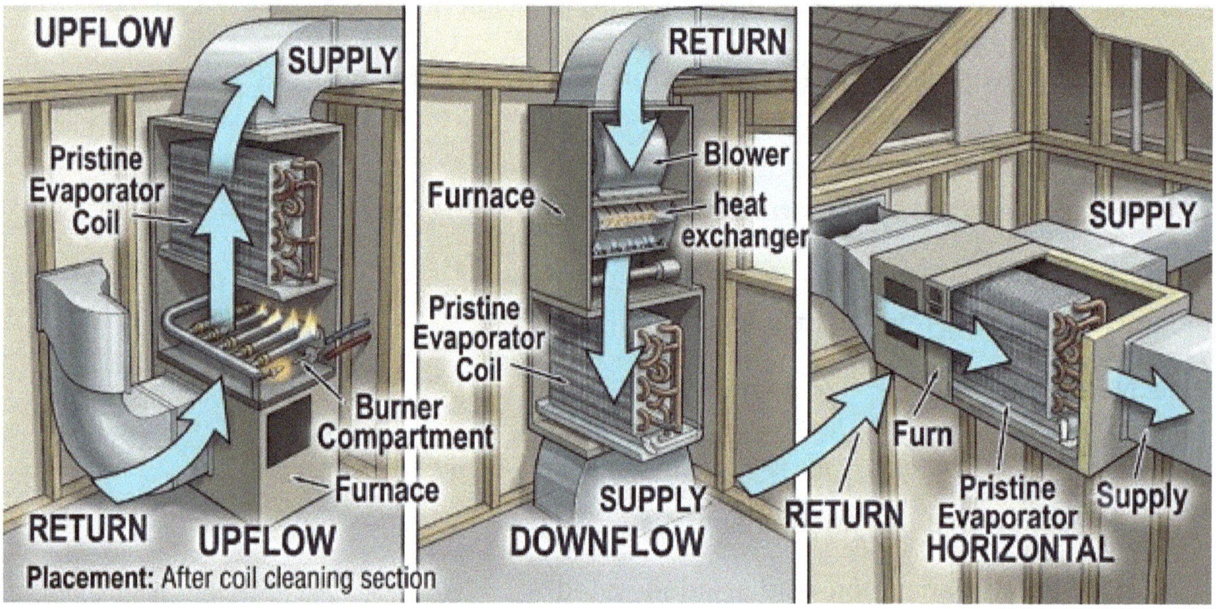

12.7 How to Clean the Drain Pan

The drain pan sits under the coil and catches condensation.

Steps:

1. Remove debris
2. Wipe with a rag
3. Spray with vinegar or mild cleaner
4. Ensure the drain hole is clear

A clogged drain pan leads to water leaks and mold.

12.8 How to Clear the Condensate Drain Line

The condensate line removes water from the drain pan.

When it clogs, the furnace or AC may shut down.

Symptoms of a clogged drain line:

- Water around furnace
- AC shuts off randomly
- Musty smell
- Gurgling sounds

How to clear it:

1. Locate the PVC drain line
2. Find the cleanout tee
3. Pour warm water or vinegar into the line
4. Use a wet/dry vacuum on the outside drain termination
5. Verify flow

Optional:

Add a condensate tablet to prevent algae growth.

12.2 Illustration — “Condensate Drain Layout”

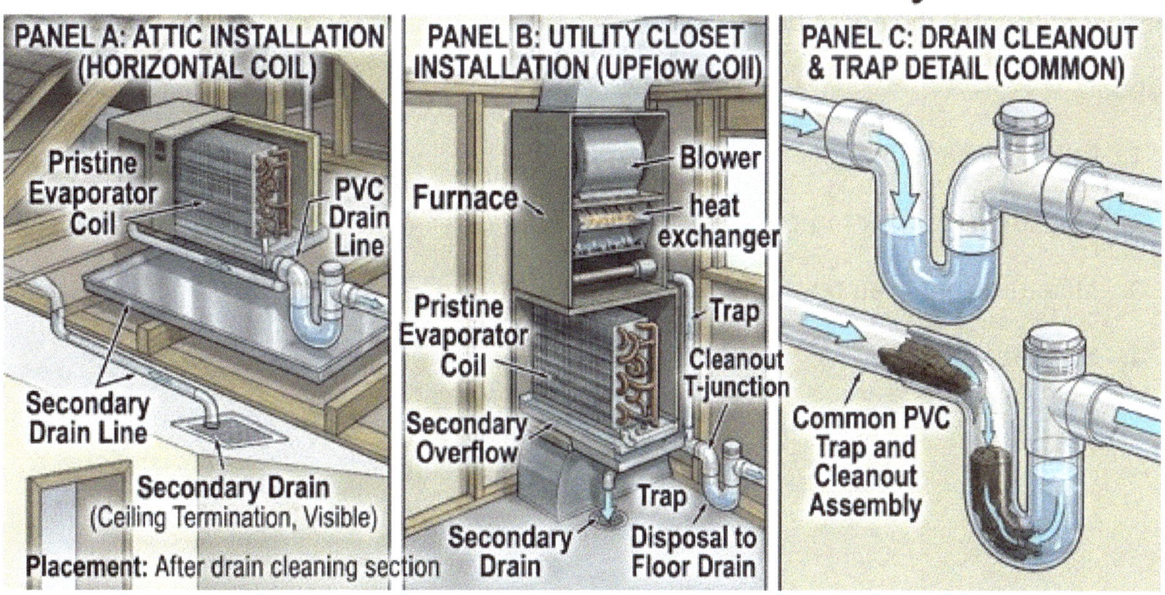

12.9 How to Prevent Coil Freeze-Ups

Freeze-ups happen when the coil gets too cold.

Causes:

- Dirty filter
- Dirty coil
- Low airflow
- Blower not running
- Low refrigerant (diagnosis only)

Prevention:

- Keep filter clean
- Keep coil clean
- Keep blower wheel clean
- Ensure return vents are open
- Maintain proper airflow

If the coil is frozen, turn off the AC and run the blower until thawed.

12.10 The Technician's Shortcut: The Temperature Drop Test

This test tells you if the coil is doing its job.

Steps:

1. Run AC for 10 minutes
2. Measure return air temperature
3. Measure supply air temperature
4. Subtract the two

Normal temperature drop:

16–22°F

If drop is too low (< 16°F):

- Dirty coil
- Low airflow
- Blower issue

If drop is too high (> 22°F):

- Coil freezing
- Airflow severely restricted

This simple test reveals a lot.

12.11 What You'll Learn Next

You've now completed the entire maintenance section of the book.

Next, we move into **Part IV — Upgrades, Thermostats & Reference Tables**, starting with:

CHAPTER 13 — Smart Thermostat Wiring & Upgrades

How to Install Modern Thermostats Without Blowing Transformers or Miswiring Your System

Smart thermostats are one of the most popular HVAC upgrades homeowners make — and one of the most commonly botched.
Not because they're complicated, but because **older HVAC systems weren't designed with smart thermostats in mind.**

The biggest issue?

Smart thermostats need constant power — and that means they need a C-wire.

This chapter teaches you how to wire, upgrade, and troubleshoot smart thermostats safely, correctly, and without damaging your furnace control board.

You'll learn:

- What the C-wire actually does
- How to identify your existing wiring
- How to install a smart thermostat
- How to handle missing C-wires
- How to use add-a-wire kits
- How to avoid blowing the transformer
- How to verify correct operation

Let's get into it.

13.1 What Makes a Thermostat "Smart"?

A smart thermostat needs:

- Constant 24V power
- Wi-Fi connectivity
- A microprocessor
- A display
- Sensors
- Software updates

All of this requires **continuous power**, not just momentary switching like old mercury thermostats.

That's why the C-wire matters.

13.2 The C-Wire: The Most Misunderstood Wire in HVAC

The **C-wire** (common wire) completes the 24V circuit and provides continuous power to the thermostat.

Without a C-wire:

- Thermostat may reboot
- Wi-Fi may drop
- Furnace may short cycle
- Transformer may overload
- Thermostat may "power steal" from other wires

With a C-wire:

- Stable power
- Stable Wi-Fi
- No flickering
- No rebooting
- No transformer stress

If you're installing a smart thermostat, the C-wire is non-negotiable.

13.3 Identifying Your Existing Thermostat Wiring

Remove the old thermostat and look at the wires.

Common wire colors:

- **R** — Red
- **C** — Blue or Black
- **W** — White
- **Y** — Yellow
- **G** — Green

But remember:

Color is a suggestion, not a guarantee.

Always verify at the furnace control board.

13.1 Illustration — "Smart Thermostat Terminal Map"

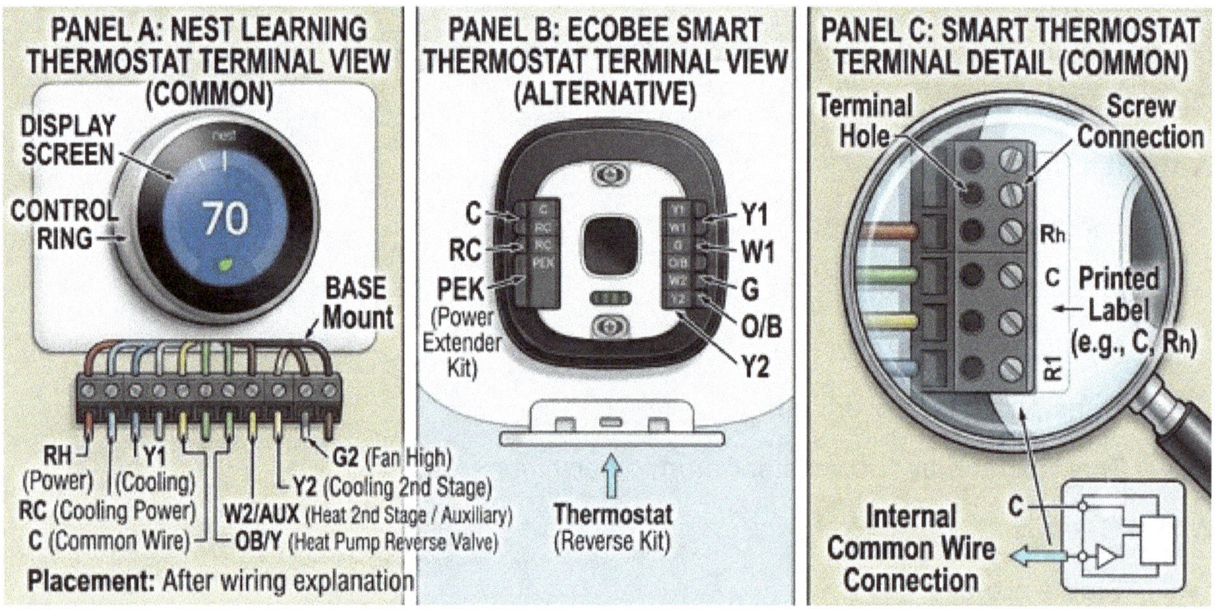

13.4 How to Verify Wiring at the Furnace Control Board

Open the blower door and locate the control board.

You'll see terminals labeled:

- R
- C
- W
- Y
- G

Match thermostat wires to board terminals.

If the thermostat has a C-wire but it's not connected at the board, connect it now.

If the thermostat doesn't have a C-wire, move to the next section.

13.5 What to Do If You Don't Have a C-Wire

You have **four** safe options.

Option 1 — Use an Add-A-Wire Kit (Most Common)

These kits use a small module to "split" one wire into two signals.

Pros:

- Works with almost any system
- Safe
- Reliable

Cons:

- Requires wiring at furnace
- Requires wiring at thermostat

This is the most common solution for older homes.

Option 2 — Run a New Thermostat Cable

If you have access to walls or attic, run a new **18/5** cable.

Pros:

- Best long-term solution
- Cleanest installation

Cons:

- Harder in finished homes

Option 3 — Use a Plug-In Transformer

Some thermostats allow external power.

Pros:

- No wiring changes
- Fast

Cons:

- Requires nearby outlet
- Visible wire

Option 4 — Use a Thermostat That Doesn't Need C-Wire

Some models "power steal" safely.

Pros:

- Easiest

Cons:

- Not compatible with all systems
- May cause furnace issues

13.6 How to Install a Smart Thermostat (Step-By-Step)

Step 1 — Kill power at furnace switch

This prevents transformer damage.

Step 2 — Remove old thermostat

Label wires as you remove them.

Step 3 — Mount new thermostat base

Ensure level and secure.

Step 4 — Connect wires to correct terminals

Match R, C, W, Y, G.

Step 5 — Restore power

Turn furnace switch back on.

Step 6 — Follow on-screen setup

Select system type:

- Gas furnace
- Heat pump
- Electric furnace
- Dual fuel

Step 7 — Test heating and cooling

Verify:

- Heat sequence
- Cool sequence
- Fan operation

13.7 How to Avoid Blowing the Transformer

This is the #1 mistake homeowners make.

NEVER:

- Short R to C
- Touch bare wires together
- Wire thermostat with power on
- Let wires fall into the wall
- Miswire heat pump systems

A short between R and C blows the 3-amp fuse — or the transformer.

13.8 How to Diagnose Thermostat Wiring Problems

Symptom: Thermostat blank

- No 24V
- Blown fuse
- Bad transformer
- Miswired C-wire

Symptom: AC turns on in heat mode

- Y and W swapped

Symptom: Blower runs constantly

- G shorted to R
- Bad fan relay

Symptom: Furnace won't ignite

- W not connected
- Broken thermostat wire

Symptom: Outdoor unit won't start

- Y not connected at board
- Broken Y wire

13.9 The Technician's Shortcut: The R-to-G Test

If thermostat is suspected bad:

- Jump **R to G** at the furnace board
- Blower should start instantly

If blower starts → thermostat or wiring is bad.
If blower doesn't → blower or board issue.

13.10 What You'll Learn Next

You've now mastered smart thermostat wiring and upgrades.

Next, we move into one of the most valuable sections of the entire book:

CHAPTER 14 — Master Troubleshooting Matrix

14.1 Illustration — "Sequence Failure Diagnostic Flowchart"

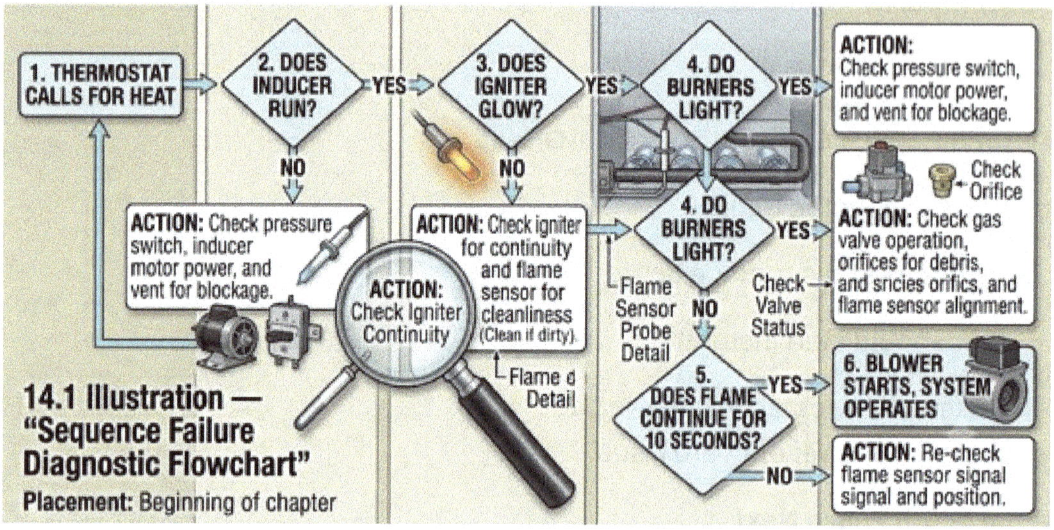

Your Quick-Reference Guide to Diagnosing Any HVAC Problem in Minutes

This chapter is the **workbench companion** for the entire book.
It's designed so you can flip straight to a symptom and immediately see:

- What's happening
- Why it's happening
- What to check first
- What to check second
- What to check last

This is the same mental flowchart technicians use on service calls — except now it's in your hands.

To keep this matrix clean and powerful, each symptom includes:

- **Primary suspects** (most common causes)
- **Secondary suspects** (less common but possible)
- **Quick tests** you can perform immediately
- **What the fix usually is**

Let's build your full diagnostic arsenal.

14.1 Furnace Troubleshooting Matrix (Heat Mode)

Symptom: Furnace won't turn on at all

Primary suspects:

- No 120V power
- Blown 3-amp fuse
- Bad transformer
- Thermostat not calling for heat

Quick tests:

- Check furnace switch
- Check breaker
- Test R → C for 24V
- Test R → W during heat call

Likely fix:
Restore power, replace fuse, fix thermostat wiring.

Symptom: Inducer runs but furnace won't ignite

Primary suspects:

- Pressure switch not closing
- Clogged inducer port
- Blocked flue
- Cracked pressure tubing

Quick tests:

- Check pressure switch continuity during inducer operation
- Inspect tubing
- Clear inducer port

Likely fix:
Clear blockage or replace pressure switch.

Symptom: Igniter glows but burners don't light

Primary suspects:

- Gas valve not opening
- No 24V to gas valve
- Dirty burners
- Low gas pressure (tech required)

Quick tests:

- Test for 24V at gas valve
- Inspect burners

Likely fix:
Clean burners or diagnose gas valve circuit.

Symptom: Burners light but shut off after 2–3 seconds

Primary suspects:

- Dirty flame sensor
- Bad ground
- Weak flame

Quick tests:

- Clean flame sensor
- Check ground wire

Likely fix:
Clean flame sensor (fixes 90% of cases).

Symptom: Furnace short cycles (starts and stops repeatedly)

Primary suspects:

- High-limit switch tripping
- Dirty filter
- Dirty blower wheel
- Dirty evaporator coil
- Closed vents

Quick tests:

- Check filter
- Check airflow
- Check temperature rise

Likely fix:
Restore airflow.

Symptom: Blower runs constantly

Primary suspects:

- Thermostat set to FAN ON
- Stuck fan relay
- High-limit switch open

Quick tests:

- Check thermostat mode
- Test G → R at board
- Check limit switch continuity

Likely fix:
Reset limit or replace fan relay.

14.2 Air Conditioner Troubleshooting Matrix (Cooling Mode)

Symptom: Outdoor unit won't turn on

Primary suspects:

- No 240V at disconnect
- Bad contactor
- No 24V to contactor coil

Quick tests:

- Test for 240V at line side
- Test for 24V at contactor coil

Likely fix:
Replace contactor or restore power.

Symptom: Outdoor fan runs but compressor won't start

Primary suspects:

- Bad capacitor (HERM side)
- Weak compressor
- Low voltage

Quick tests:

- Test capacitor μF
- Listen for compressor hum

Likely fix:
Replace capacitor.

Symptom: Compressor hums but won't start

Primary suspects:

- Dead capacitor
- Locked compressor
- Low voltage

Quick tests:

- Test capacitor
- Check voltage at contactor

Likely fix:
Replace capacitor or diagnose compressor.

Symptom: Outdoor unit runs but air is warm inside

Primary suspects:

- Dirty condenser coil
- Dirty evaporator coil
- Low airflow
- Low refrigerant (diagnosis only)

Quick tests:

- Check condenser coil
- Check filter
- Check temperature drop

Likely fix:
Clean coils and restore airflow.

Symptom: AC freezes up

Primary suspects:

- Dirty filter
- Dirty evaporator coil
- Blower not running
- Low airflow
- Low refrigerant (diagnosis only)

Quick tests:

- Inspect coil
- Check blower
- Check airflow

Likely fix:
Thaw coil and restore airflow.

Symptom: AC runs constantly

Primary suspects:

- Dirty condenser coil
- Undersized system
- Low refrigerant (diagnosis only)
- Poor airflow

Quick tests:

- Check condenser coil
- Check temperature drop

Likely fix:
Clean condenser coil.

14.3 Blower Motor Troubleshooting Matrix

Symptom: Blower won't start

Primary suspects:

- Bad capacitor
- Bad motor
- Bad relay
- No 120V to motor

Quick tests:

- Test capacitor
- Test for 120V at motor leads

Likely fix:
Replace capacitor or motor.

Symptom: Blower is loud or rattling

Primary suspects:

- Dirty blower wheel
- Loose set screw
- Bent wheel

Quick tests:

- Inspect wheel
- Check motor mount

Likely fix:
Clean or tighten blower assembly.

14.4 Thermostat Troubleshooting Matrix

Symptom: Thermostat blank

Primary suspects:

- No 24V
- Blown fuse
- Bad transformer
- Miswired C-wire

Quick tests:

- Test R → C
- Check fuse

Likely fix:
Restore 24V power.

Symptom: AC turns on in heat mode

Primary suspects:

- Y and W swapped

Quick tests:

- Verify wiring at thermostat and board

Likely fix:
Correct wiring.

14.5 The Ultimate Quick-Check Flowchart (Text Version)

Is there power?

↓

Is thermostat sending signal?

↓

Is control board responding?

↓

Is sequence of operation starting?

↓

Where does the sequence stop?

↓

Test that component.

↓

Fix airflow, electrical, or safety issue.

This is the technician's entire diagnostic method in one flow.

14.6 What You'll Learn Next

You've now completed the most powerful diagnostic section of the book.

Next, we finish the book with:

CHAPTER 15 — Glossary & Reference Tables

15.1 Illustration — “HVAC System Overview Diagram”

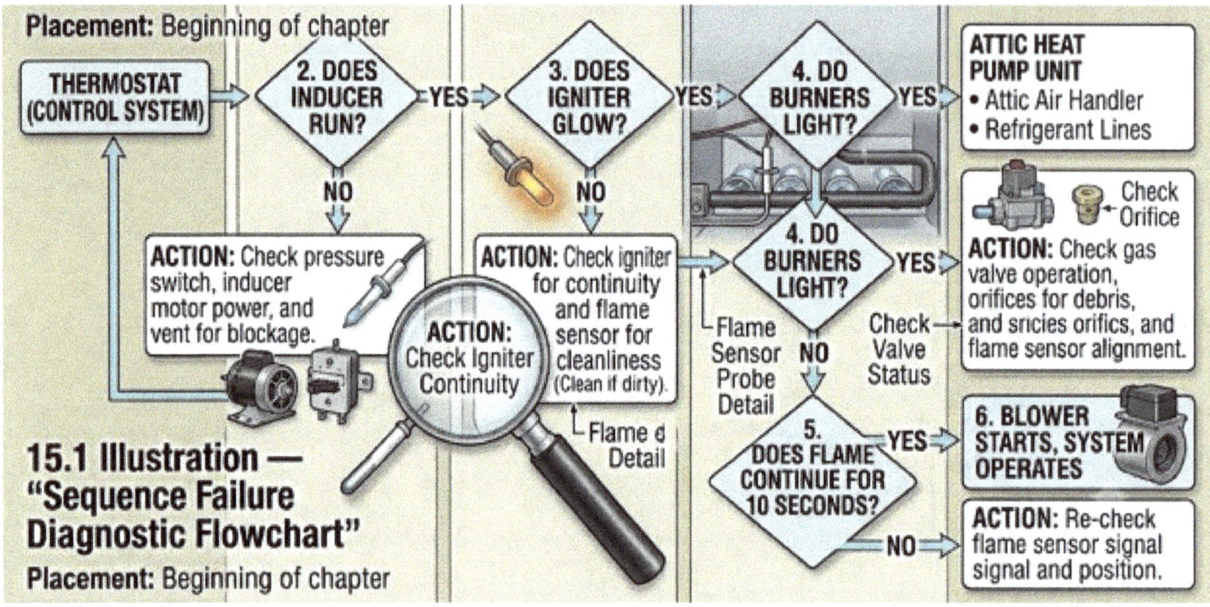

Your Complete Quick-Reference Section for HVAC Terms, Components, Wiring, and System Specs

This final chapter is designed to be the **back-of-the-book toolbox** you flip to whenever you need a fast definition, a wiring reminder, a component explanation, or a quick-reference chart.

It’s clean, simple, and built for real-world use — whether you’re diagnosing a furnace at midnight or wiring a thermostat on a Saturday morning.

Let’s finish strong.

15.1 Glossary of Essential HVAC Terms

A-Coil

An evaporator coil shaped like the letter “A,” mounted above the furnace.

Air Handler

Indoor unit used in heat pump or electric systems; contains blower and coil.

Blower Motor

Moves air through the furnace and duct system.

BTU (British Thermal Unit)

Measurement of heating or cooling capacity.

Capacitor

Stores electrical energy to help motors start and run.

C-Wire (Common Wire)

Provides continuous 24V power to thermostats.

Compressor

The heart of the AC system; pumps refrigerant.

Condenser Coil

Outdoor coil that releases heat to the outside air.

Contactor

Electrically controlled switch that sends 240V to compressor and fan.

Control Board

The furnace's "brain" that manages the sequence of operation.

Evaporator Coil

Indoor coil that absorbs heat from the home.

Flame Sensor

Detects flame during ignition; shuts off gas if flame is not present.

Heat Exchanger

Transfers heat from combustion gases to indoor air.

High-Limit Switch

Safety device that shuts off burners if furnace overheats.

HSI (Hot Surface Igniter)

Glows red-hot to ignite gas burners.

Inducer Motor

Creates draft for safe combustion.

MERV Rating

Filter efficiency rating; higher = finer filtration.

Pressure Switch

Verifies inducer draft before ignition.

Refrigerant

Chemical that absorbs and releases heat in AC systems.

Return Air

Air pulled from the home back to the furnace.

Supply Air

Conditioned air delivered to the home.

Thermistor

Temperature-sensitive resistor used for sensing.

Transformer

Converts 120V to 24V for control circuits.

15.2 Thermostat Wiring Reference Table

Terminal	Color (Typical)	Function
R	Red	24V power
C	Blue/Black	Common
W	White	Heat
Y	Yellow	Cooling
G	Green	Fan
O/B	Orange	Heat pump reversing valve

Terminal	Color (Typical)	Function
Aux/E	Brown	Auxiliary heat

Note: Colors vary — always verify at the control board.

15.3 Furnace Sequence of Operation (Quick Reference)

1. Thermostat calls for heat (W → R)
2. Inducer motor starts
3. Pressure switch closes
4. Igniter glows
5. Gas valve opens
6. Burners ignite
7. Flame sensor verifies flame
8. Blower motor starts

If the sequence stops, diagnose that step.

15.4 AC Cooling Sequence (Quick Reference)

1. Thermostat calls for cooling (Y → R)
2. Blower motor starts
3. Contactor pulls in
4. Compressor starts
5. Condenser fan starts
6. Evaporator coil absorbs heat
7. Condenser coil rejects heat

If outdoor unit doesn't run → check contactor and capacitor.

15.5 Capacitor Testing Reference

Dual-Run Capacitor Terminals:

- **C** — Common
- **FAN** — Fan motor
- **HERM** — Compressor

Tolerance Rule:

Must be within **±6%** of rated μF.

Common Failures:

- Bulging top
- Leaking oil
- Low μF reading
- Zero reading

15.6 Temperature Rise & Drop Reference

Heating Mode (Temperature Rise)

Normal: **40–70°F**
High rise → airflow problem.

Cooling Mode (Temperature Drop)

Normal: **16–22°F**
Low drop → airflow issue.
High drop → coil freezing.

15.7 Common Component Voltage Reference

Component	Voltage
Thermostat circuits	24V AC
Gas valve	24V AC
Contactor coil	24V AC
Blower motor	120V AC
Inducer motor	120V AC
Igniter	120V AC
Compressor	240V AC
Condenser fan	240V AC

15.8 Safety Switch Reference Table

Device	Normal State	Purpose
High-Limit Switch	Closed	Prevent overheating
Rollout Switch	Closed	Prevent flame rollout
Pressure Switch	Open (closes during inducer)	Verify draft
Flame Sensor	N/A	Verify flame presence

15.9 Common Furnace Flash Codes (Generalized)

Flashes	Meaning
1	Ignition failure
2	Pressure switch stuck open
3	Pressure switch stuck closed
4	High-limit switch open
5	Flame sensed with no call
Steady On	Normal operation

Always check the furnace's specific sticker.

15.10 Motor Capacities & Wire Gauge Reference

Typical Blower Motor Capacitors:

- 5 μF
- 7.5 μF
- 10 μF

Typical Condenser Capacitors:

- 35/5 μF
- 40/5 μF
- 45/5 μF

Thermostat Wire Gauge:

- 18 AWG (standard)

15.11 Efficiency Ratings Reference

AFUE (Furnace Efficiency)

- 80% = standard
- 90–97% = high efficiency

SEER (Cooling Efficiency)

- 13–14 SEER = older systems
- 15–18 SEER = modern systems
- 19+ SEER = high efficiency

15.12 Final Notes for the Reader

You now have:

- The sequences
- The diagnostics
- The wiring
- The maintenance
- The safety devices
- The troubleshooting matrix
- The reference tables

This book gives you the same foundation technicians use every day.

You've built a complete HVAC knowledge base — and you did it step by step.

CHAPTER 16 — Air-Source Heat Pumps: Operation, Components & Diagnostics

Heat pumps are one of the most misunderstood systems in residential HVAC. Homeowners think they are complicated. Technicians know the truth: a heat pump is simply an air conditioner that can run in reverse.

If you understand cooling mode, you already understand 70% of how a heat pump works. This chapter teaches the remaining 30% — the reversing valve, the defrost cycle, auxiliary heat, and the heat-pump-specific diagnostic logic that separates a beginner from a technician.

16.1 What a Heat Pump Really Is

A heat pump is a refrigeration system capable of moving heat in two directions:

- **Cooling Mode:** Moves heat from indoors to outdoors
- **Heating Mode:** Moves heat from outdoors to indoors

The system does not "create" heat. It relocates it. Even cold outdoor air contains usable heat energy.

A heat pump is built from the same components as a standard AC system:

- Compressor
- Outdoor fan motor
- Dual-run capacitor
- Contactor
- Evaporator coil
- Condenser coil
- 24V control system

The only major addition is the **reversing valve**, which changes the direction of refrigerant flow.

16.2 The Reversing Valve (The Heart of Heat Pump Operation)

The reversing valve is a four-way directional valve located in the outdoor unit. It determines whether the system is in heating or cooling mode.

How It Works

- In **cooling mode**, the indoor coil is the evaporator and the outdoor coil is the condenser.
- In **heating mode**, the reversing valve shifts, making the indoor coil the condenser and the outdoor coil the evaporator.

O/B Logic

Thermostats energize the reversing valve using one of two conventions:

- **O-energized systems:**
  The reversing valve is energized in *cooling*. (Most common: Rheem, Ruud, Goodman, Amana.)
- **B-energized systems:**
  The reversing valve is energized in *heating*. (Common in older systems.)

Diagnostic Rule

If the system heats when it should cool — or cools when it should heat — the reversing valve or its 24V control signal is the first suspect.

CHAPTER 16 — Heat Pumps

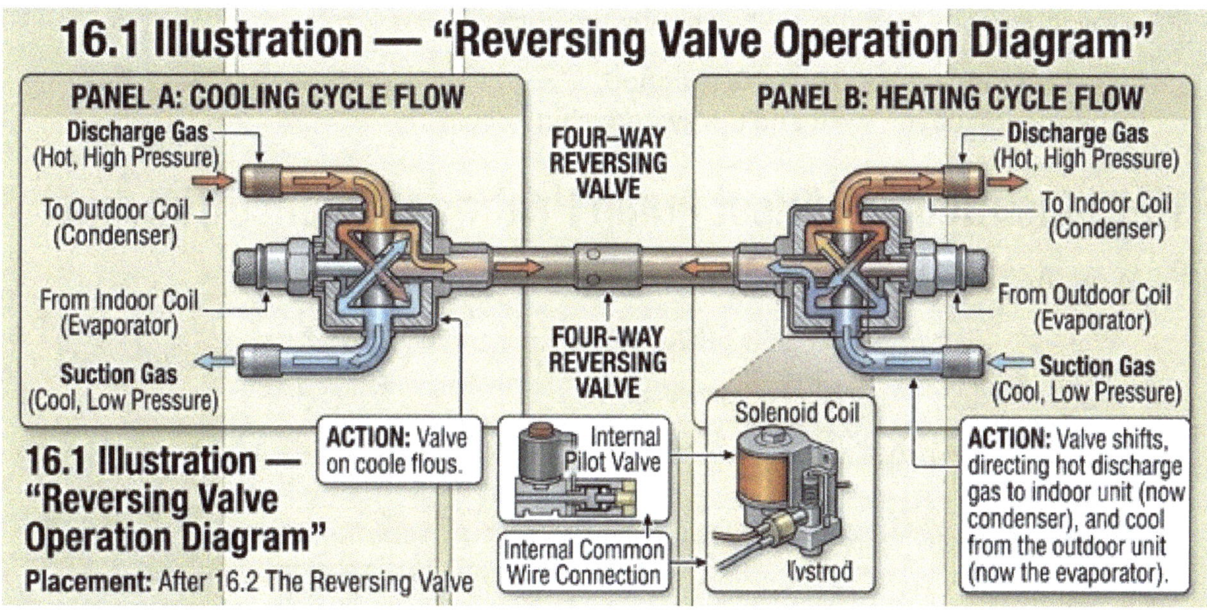

16.3 Heat Pump Heating Sequence of Operation

Heating mode follows a strict electrical and mechanical order. Understanding this sequence allows you to diagnose failures without guessing.

Step-By-Step Heating Sequence

1. **Thermostat calls for heat (Y + O/B + G).**
   The thermostat sends 24V to the outdoor unit and blower.
2. **Indoor blower starts.**
   Warm air must move across the indoor coil (now acting as a condenser).
3. **Outdoor contactor pulls in.**
   24V energizes the contactor coil.
4. **Compressor starts.**
   Refrigerant begins flowing in the heating direction.
5. **Outdoor fan starts.**
   The outdoor coil is now the evaporator, absorbing heat from outside air.
6. **Reversing valve shifts.**
   The valve energizes or de-energizes depending on O/B logic.
7. **Indoor coil releases heat.**
   The indoor coil becomes hot, warming the home.
8. **System runs until setpoint is reached.**
   Thermostat opens Y → R and the system shuts down.

16.2 Illustration — "Heat Pump Heating Sequence Timeline"

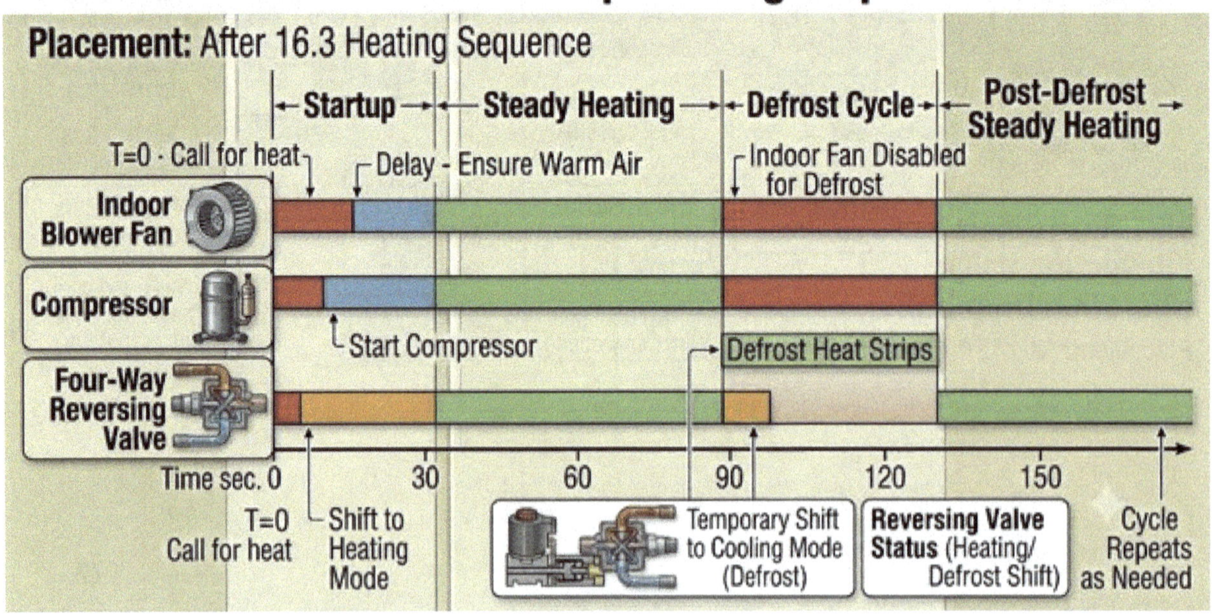

16.4 Auxiliary Heat (Electric Heat Strips)

Heat pumps lose efficiency as outdoor temperatures drop. Below ~35°F, the system may require supplemental heat.

Auxiliary Heat Activates When:

- Outdoor temperature is too low
- Thermostat detects long runtimes
- Defrost cycle temporarily reverses into cooling
- Emergency heat mode is selected

What Auxiliary Heat Is

Electric resistance heat strips inside the air handler. They operate like a giant toaster coil.

Diagnostic Rule

If a heat pump blows cool air in winter, but warms up when "EM HEAT" is selected, the issue is almost always:

- A failed outdoor unit
- A failed reversing valve
- A failed defrost board
- A failed outdoor sensor

16.5 The Defrost Cycle (The Most Misunderstood Part of Heat Pumps)

In heating mode, the outdoor coil becomes cold — cold enough to freeze moisture from the air. Frost buildup reduces airflow and efficiency.

How Defrost Works

1. Outdoor coil temperature sensor detects frost
2. Defrost board initiates a timed or demand-based defrost
3. System temporarily switches into cooling mode
4. Outdoor fan shuts off
5. Indoor auxiliary heat turns on
6. Frost melts off the outdoor coil
7. System returns to heating mode

Normal Symptoms During Defrost

- Steam rising from outdoor unit
- Outdoor fan stopping
- Loud “whoosh” as reversing valve shifts
- Temporary cool air indoors (offset by heat strips)

These are normal and not signs of failure.

16.3 Illustration — “Defrost Cycle Diagram”

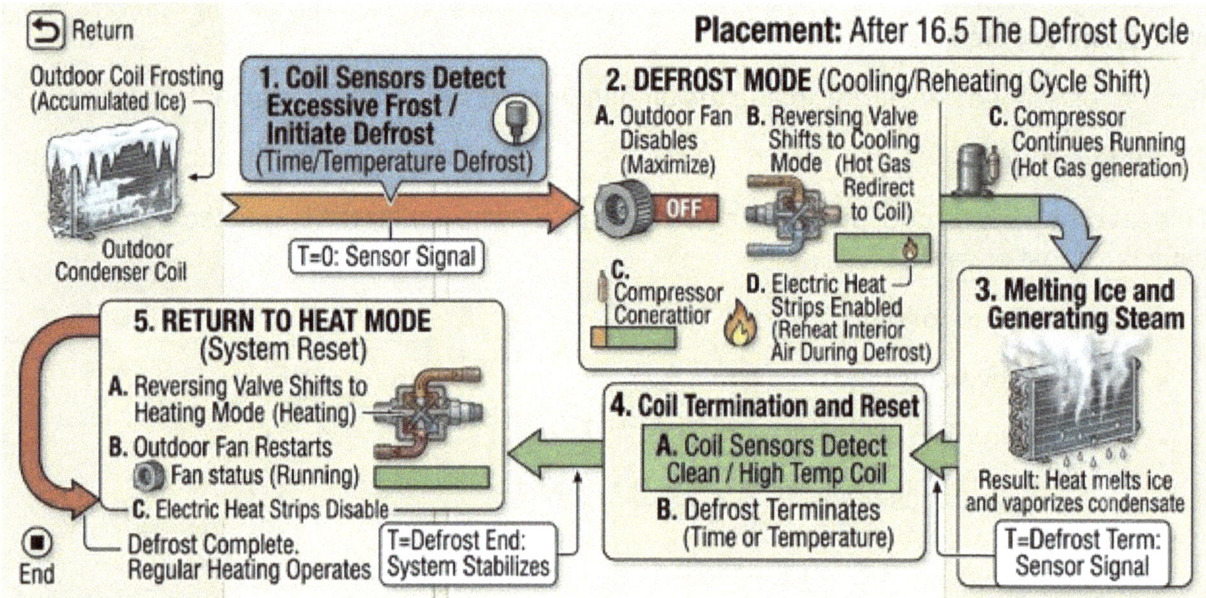

16.6 Common Heat Pump Failures

1. Reversing Valve Stuck

Symptoms:

- System cools when calling for heat
- System heats when calling for cool
- Loud buzzing at valve coil

2. Defrost Board Failure

Symptoms:

- Outdoor unit freezes into a solid block of ice
- No defrost cycles
- Outdoor fan never shuts off

3. Outdoor Thermistor Failure

Symptoms:

- Defrost too often
- Defrost never occurs
- System short cycles

4. Auxiliary Heat Failure

Symptoms:

- Heat pump runs constantly
- Lukewarm air in cold weather
- Emergency heat works, but normal heat does not

5. Low Airflow

Symptoms:

- Indoor coil overheating
- High electric bills
- Poor heating performance

16.7 Heat Pump Troubleshooting Matrix

Symptom: System blows cold air in heat mode
Likely Causes:

- Reversing valve stuck
- Defrost board stuck in cooling
- Outdoor unit not running
- Low refrigerant (diagnosis only)

Symptom: Outdoor unit frozen solid
Likely Causes:

- Defrost board failure
- Outdoor sensor failure
- Outdoor fan failure
- Low airflow across outdoor coil

Symptom: Heat pump runs constantly
Likely Causes:

- Low outdoor temperature
- Auxiliary heat not engaging
- Undersized system
- Dirty indoor coil or filter

Symptom: Loud whooshing sound
Likely Cause:

- Reversing valve shifting (normal)

16.8 What You'll Learn in Chapter 17

Chapter 17 will teach you:

- How to test a reversing valve coil
- How to test defrost board inputs and outputs
- How to test outdoor thermistors
- How to diagnose auxiliary heat failures
- How to identify heat-pump-specific electrical problems
- How to separate refrigerant-side issues from electrical issues (legally and safely)

This is where heat pump diagnostics become simple, predictable, and technician-level.

CHAPTER 17 — Heat Pump Diagnostics & Field Testing

Heat pumps introduce a handful of components that standard air conditioners do not have: the reversing valve, the defrost board, outdoor thermistors, and auxiliary heat controls. Once you understand how to test these components, heat pump diagnostics become just as predictable as cooling mode.

This chapter teaches you how to test every heat-pump-specific component safely, legally, and with technician-level accuracy.

17.1 The Heat Pump Diagnostic Mindset

Heat pump troubleshooting is built on three principles:

1. **Follow the sequence of operation**
   Heating mode has a strict electrical order. Find the step where the chain stops.
2. **Separate electrical failures from refrigerant failures**
   You can legally diagnose electrical and airflow issues without touching refrigerant.
3. **Test components, don't guess**
   Every heat pump component can be tested with a multimeter.

If you can test voltage, continuity, and resistance, you can diagnose 90% of heat pump failures.

17.2 Testing the Reversing Valve Coil

The reversing valve coil is a simple 24-volt solenoid. If it fails, the system may:

- Cool when calling for heat
- Heat when calling for cool
- Fail to shift modes
- Buzz loudly during operation

How to Test the Reversing Valve Coil

1. **Turn off power at the disconnect.**
2. **Locate the reversing valve coil** on the outdoor unit.
3. **Disconnect the two low-voltage wires** feeding the coil.
4. **Set your meter to resistance (ohms).**
5. **Measure coil resistance.**

Typical readings:

- 10–40 ohms = normal
- Infinite resistance = open coil (failed)
- 0–1 ohm = shorted coil (failed)

Voltage Test (During Operation)

1. Restore power.
2. Call for cooling (O-energized systems) or heating (B-energized systems).
3. Measure voltage across the coil terminals.

Expected reading: **24 VAC**.

If you have 24V but the valve does not shift, the coil is bad.
If you have **no 24V**, the issue is upstream:

- Thermostat
- Control board
- Wiring
- Defrost board

CHAPTER 17 — Heat Pump Diagnostics

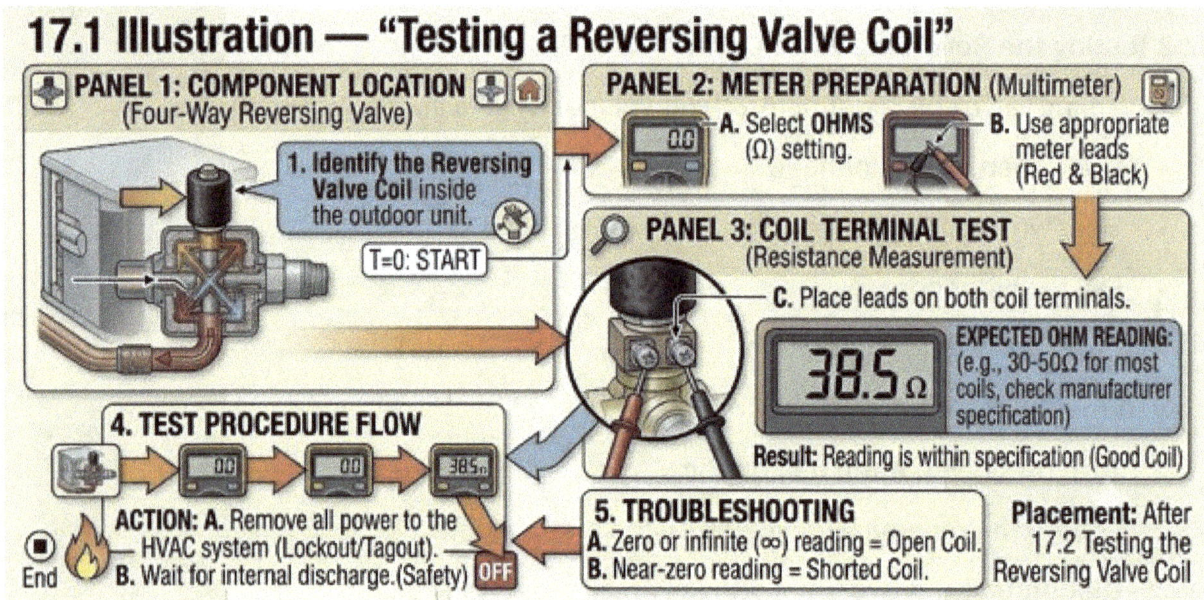

17.3 Testing the Defrost Board

The defrost board is the “brain” of the outdoor unit in heating mode. It controls:

- Defrost initiation
- Reversing valve timing
- Outdoor fan shutoff
- Auxiliary heat activation
- Compressor protection

Symptoms of a Bad Defrost Board

- Outdoor unit freezes into a block of ice
- Defrost never occurs
- Defrost occurs too often
- Outdoor fan never shuts off
- System short cycles
- Reversing valve does not shift during defrost

Defrost Board Input Tests

With the system running in heat mode:

1. **Check for 24V between R and C** on the board.
   Confirms the board is powered.
2. **Check for 24V from Y to C.**
   Confirms the thermostat is calling for heat.
3. **Check the outdoor thermistor resistance** (see next section).
   Confirms the board is receiving temperature data.

Defrost Board Output Tests

1. **Check for 24V to the reversing valve coil** during defrost.
2. **Check for 24V to the outdoor fan relay** (fan should shut off during defrost).
3. **Check for 24V to auxiliary heat** (indoor strips should energize).

If the board receives correct inputs but sends no outputs, the board has failed.

CHAPTER 17 — Heat Pump Diagnostics

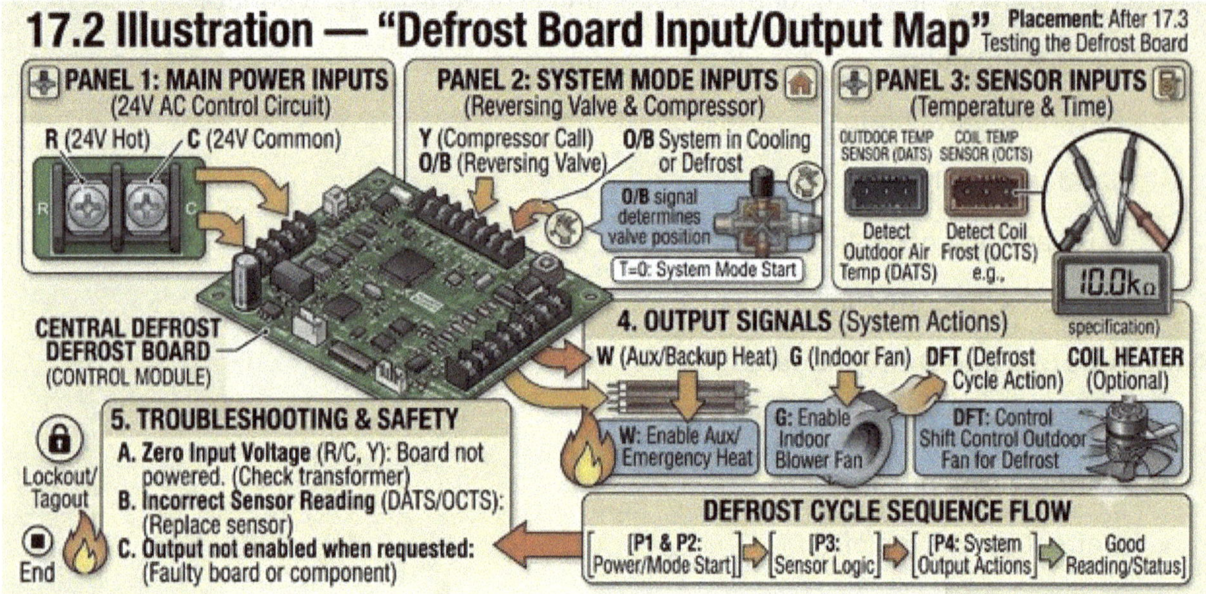

17.4 Testing Outdoor Thermistors (Temperature Sensors)

Outdoor thermistors tell the defrost board when frost is forming. A failed thermistor can cause:

- No defrost
- Constant defrost
- Short cycling
- Poor heating performance

How to Test a Thermistor

1. Turn off power.
2. Disconnect the thermistor from the board.
3. Set your meter to resistance (ohms).
4. Measure resistance across the thermistor leads.

Thermistors are temperature-dependent resistors.
As temperature drops, resistance rises.

Typical values (approximate):

- 70°F: 10k ohms
- 40°F: 20k ohms
- 20°F: 30k–40k ohms

If the reading is:

- Infinite = open thermistor (failed)
- 0–1 ohm = shorted thermistor (failed)
- Wildly outside expected range = failed

17.5 Testing Auxiliary Heat (Electric Heat Strips)

Auxiliary heat is critical in cold weather. If it fails, the home will never reach setpoint.

Symptoms of Failed Auxiliary Heat

- Heat pump runs constantly
- Lukewarm air from vents
- Thermostat displays "AUX HEAT" but no temperature rise
- Emergency heat works, but normal heat does not

How to Test Heat Strips

1. Turn off power.
2. Remove the air handler panel.
3. Locate the heat strip sequencers or relays.
4. Set your meter to continuity.
5. Test each heating element for continuity.

Expected reading: **Low resistance (1–20 ohms)**.

If the element is open (infinite resistance), it has failed.

Voltage Test (During Operation)

1. Restore power.
2. Call for heat with a large temperature increase (5–10°F).
3. Measure voltage across the heat strip terminals.

Expected reading: **240 VAC**.

If no voltage is present:

- Sequencer failed
- Relay failed
- Thermostat not sending W2
- Control board not energizing auxiliary heat

17.6 Testing the Outdoor Fan Motor (Heat Pump Mode)

In heating mode, the outdoor fan must run continuously except during defrost.

Symptoms of Outdoor Fan Failure

- Outdoor coil freezes
- High head pressure
- Compressor overheating
- Loud humming from outdoor unit

How to Test

1. Test capacitor (if PSC motor).
2. Test motor windings for continuity.
3. Test for 240V at the motor during operation.

If the motor has power but does not run, it has failed.

17.7 Testing the Compressor in Heat Pump Mode

The compressor is identical to an AC compressor.
Testing is the same:

- Test capacitor (HERM terminal)
- Test for 240V at the contactor load side
- Test for voltage drop under load
- Test windings for continuity

If the compressor hums but does not start, the capacitor is the most common failure.

17.8 Heat Pump Diagnostic Flowchart

Symptom: No heat, outdoor unit not running
Check:

- 24V at contactor
- Defrost board power
- Thermostat O/B signal

Symptom: Outdoor unit frozen solid
Check:

- Defrost board outputs
- Outdoor thermistor
- Outdoor fan motor

Symptom: System heats in cooling mode
Check:

- Reversing valve coil
- O/B thermostat setting
- Defrost board output

Symptom: Heat pump runs constantly
Check:

- Auxiliary heat operation
- Airflow
- Outdoor temperature
- Thermostat staging

17.9 What You'll Learn in Chapter 18

Chapter 18 introduces **geothermal (ground-source) heat pumps**, including:

- Loop field types
- Circulator pumps
- Antifreeze mixtures
- Geothermal heating and cooling sequence
- Geothermal thermostat logic
- Common geothermal failures

This is where your book expands into advanced HVAC territory while remaining safe, legal, and beginner-friendly.

CHAPTER 18 — Geothermal (Ground-Source) Heat Pumps

Geothermal systems are the most efficient residential HVAC systems ever created. They use the stable temperature of the earth as their heat source and heat sink, allowing them to heat and cool a home with far less energy than any air-source heat pump or gas furnace.

Despite their reputation for complexity, geothermal systems follow the same basic logic as air-source heat pumps. The difference lies in how they exchange heat with the environment. Instead of using outdoor air, they use the ground.

This chapter teaches you how geothermal systems work, the components they use, the types of loop fields, and the sequence of operation for both heating and cooling.

18.1 What a Geothermal Heat Pump Really Is

A geothermal heat pump is a **water-source heat pump** connected to a buried loop of piping filled with antifreeze solution. The loop circulates fluid through the earth, where temperatures remain stable year-round (typically 45–60°F depending on region).

Key Differences from Air-Source Heat Pumps

- No outdoor fan motor
- No outdoor coil
- No defrost cycle
- No exposure to outdoor weather
- Higher efficiency (COP 3–5+)
- Longer equipment lifespan

Key Similarities

- Same compressor types
- Same reversing valve
- Same 24V control logic
- Same thermostat wiring
- Same auxiliary heat options

If you understand air-source heat pumps, geothermal is simply the next step.

18.2 Closed-Loop vs. Open-Loop Systems

Geothermal systems come in two major categories: **closed-loop** and **open-loop**.

Closed-Loop Systems

A sealed loop of pipe circulates antifreeze solution through the ground.

Types of closed loops:

- **Horizontal loops** (trenches 4–6 feet deep)
- **Vertical loops** (boreholes 150–300 feet deep)
- **Pond/lake loops** (coils sunk in a body of water)

Advantages:

- Low maintenance
- Long lifespan (50+ years)
- No water quality issues

Open-Loop Systems

Use groundwater directly from a well.

Advantages:

- Extremely efficient
- Lower installation cost

Disadvantages:

- Water quality issues
- Scaling and mineral buildup
- Requires discharge location

Diagnostic Rule

If the system uses a **flow center** with circulator pumps, it is closed-loop.
If it uses a **well pump**, it is open-loop.

CHAPTER 18 — Geothermal Systems

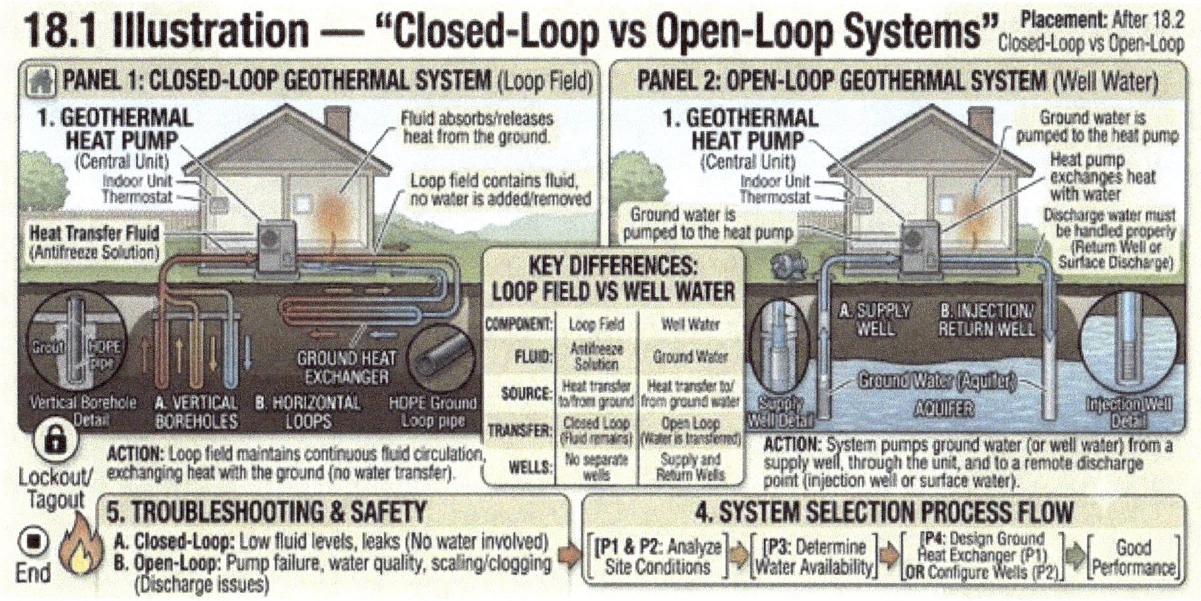

18.3 The Loop Field: How the Earth Stores Heat

The loop field is the geothermal equivalent of the outdoor coil in an air-source heat pump.

In Heating Mode

The loop absorbs heat from the earth and carries it to the heat pump.

In Cooling Mode

The loop rejects heat into the earth.

Because the earth stays at a stable temperature, geothermal systems avoid the efficiency losses that air-source heat pumps experience in extreme weather.

CHAPTER 18 — Geothermal Systems

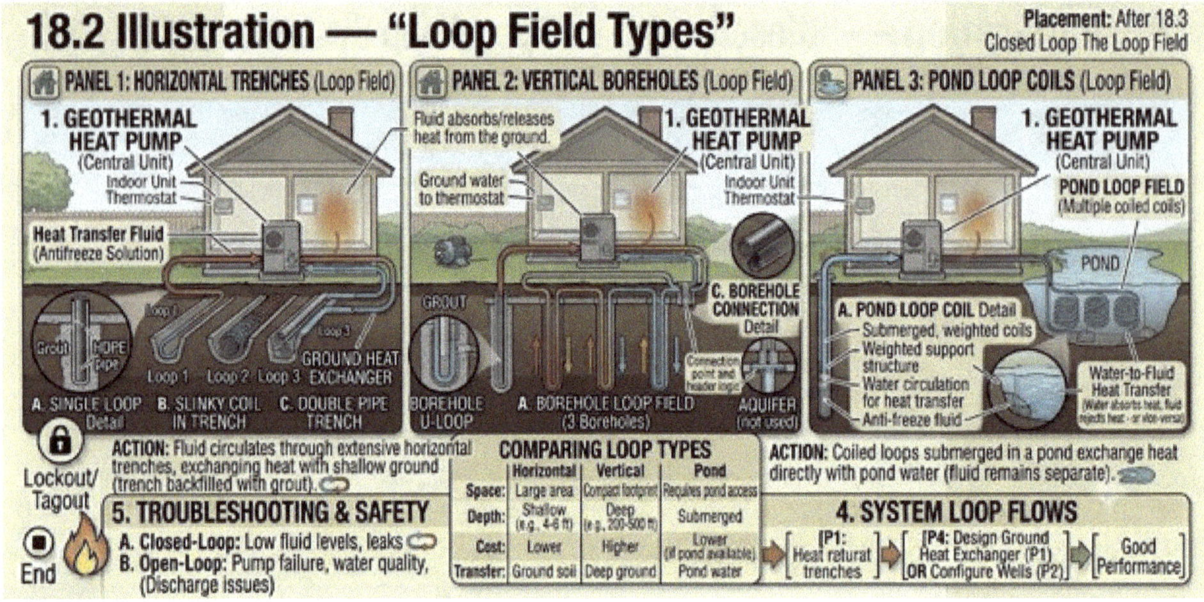

18.4 Circulator Pumps and Flow Centers

Closed-loop systems use **circulator pumps** to move antifreeze solution through the loop field.

Flow Center Components

- One or two circulator pumps
- Check valves
- Air elimination chamber
- Isolation valves
- Pressure/temperature ports

Symptoms of Flow Problems

- Poor heating or cooling
- High loop temperature (cooling mode)
- Low loop temperature (heating mode)
- Compressor lockouts
- Gurgling or air noise in the loop

Common Causes

- Air in the loop
- Failed circulator pump
- Low loop pressure
- Blocked loop field

CHAPTER 18 — Geothermal Systems

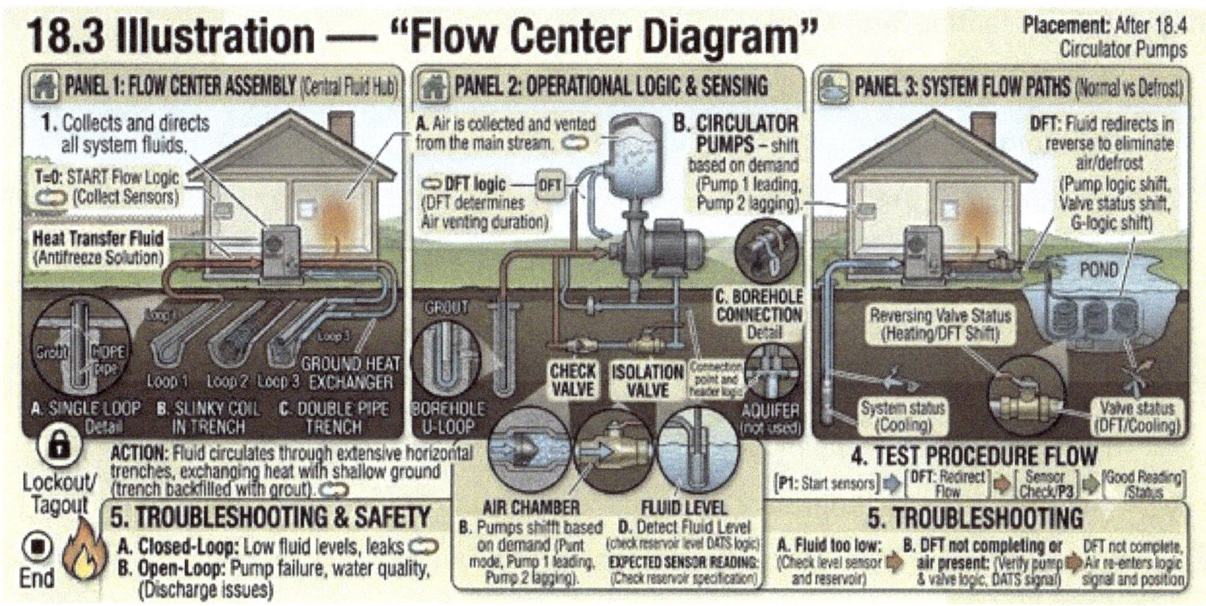

18.5 Antifreeze Mixtures

Closed-loop systems use antifreeze to prevent freezing in winter.

Common fluids:

- Propylene glycol
- Ethanol
- Methanol

Why Antifreeze Matters

- Prevents freezing in the loop
- Improves heat transfer
- Protects against corrosion

Diagnostic Rule

If loop temperature drops below 30°F in heating mode, the antifreeze concentration may be incorrect.

18.6 Geothermal Heating Sequence of Operation

The geothermal heating sequence is similar to an air-source heat pump, but without outdoor fan or defrost logic.

Step-By-Step Heating Sequence

1. Thermostat calls for heat (Y + G).
2. Circulator pump starts.
3. Compressor starts.
4. Reversing valve shifts to heating mode.
5. Indoor coil (condenser) releases heat.
6. Auxiliary heat activates if needed.
7. System runs until setpoint is reached.

Key Differences from Air-Source

- No outdoor fan
- No defrost cycle
- Loop temperature determines efficiency

18.7 Geothermal Cooling Sequence of Operation

Cooling mode is nearly identical to an air-source heat pump.

Step-By-Step Cooling Sequence

1. Thermostat calls for cooling (Y + G).
2. Circulator pump starts.
3. Compressor starts.
4. Reversing valve shifts to cooling mode.
5. Indoor coil (evaporator) absorbs heat.
6. Loop field rejects heat into the earth.
7. System runs until setpoint is reached.

Diagnostic Rule

If loop temperature rises excessively in cooling mode, the loop field may be undersized or circulation may be restricted.

18.8 Common Geothermal Failures

1. Circulator Pump Failure

Symptoms:

- No heating or cooling
- Compressor short cycling
- High loop temperature (cooling)
- Low loop temperature (heating)

2. Air in the Loop

Symptoms:

- Gurgling sounds
- Poor heat transfer
- Flow center cavitation

3. Low Loop Pressure

Symptoms:

- Lockouts
- Poor performance
- Flow center noise

4. Scaling (Open-Loop Systems)

Symptoms:

- Reduced water flow
- High head pressure
- Poor cooling performance

5. Thermistor Failure

Symptoms:

- Incorrect loop temperature readings
- Improper staging
- Lockouts

18.9 Geothermal Troubleshooting Matrix

Symptom: Poor heating performance
Likely Causes:

- Low loop temperature
- Failed circulator pump
- Air in loop
- Undersized loop field

Symptom: Poor cooling performance
Likely Causes:

- High loop temperature
- Flow restriction
- Scaling (open-loop)

Symptom: System locks out
Likely Causes:

- Low flow
- Low pressure
- Thermistor failure
- Compressor overload

Symptom: Gurgling or bubbling noises
Likely Cause:

- Air in loop

18.10 What You'll Learn in Chapter 19

Chapter 19 will teach you:

- How to test loop pressure
- How to test circulator pumps
- How to diagnose air in the loop
- How to diagnose flow center issues
- How to identify open-loop water quality problems
- How to use temperature and pressure readings to pinpoint failures

This is where geothermal diagnostics become predictable and technician-level.

CHAPTER 19 — Geothermal Diagnostics & Maintenance

Geothermal systems are extremely reliable, but when they fail, the symptoms can be confusing for anyone who doesn't understand loop dynamics, circulator pumps, and water-source heat transfer. The good news is that geothermal diagnostics follow the same logic as air-source heat pumps: verify power, verify flow, verify sequence.

This chapter teaches you how to diagnose geothermal systems safely, legally, and accurately — without touching refrigerant or performing EPA-restricted work.

19.1 The Geothermal Diagnostic Mindset

Geothermal troubleshooting is built on four principles:

1. **Flow is everything**
   If the loop fluid isn't circulating properly, the system cannot heat or cool.
2. **Loop temperature tells the story**
   High loop temps in cooling or low loop temps in heating point directly to flow issues.
3. **Electrical diagnostics are identical to air-source heat pumps**
   Same compressor tests, same reversing valve tests, same thermostat logic.
4. **Most failures occur outside the heat pump itself**
   Flow centers, pumps, loop pressure, and water quality cause the majority of issues.

19.2 Testing Loop Pressure

Closed-loop systems operate under a specific pressure range, typically **30–60 PSI** depending on loop depth and fluid type.

Symptoms of Low Loop Pressure

- Poor heating performance
- Poor cooling performance
- Flow center cavitation
- Gurgling or bubbling noises
- Compressor lockouts

How to Test Loop Pressure

1. Turn off the system.
2. Locate the pressure/temperature (P/T) ports on the flow center.
3. Insert a P/T needle into the pressure port.
4. Read system pressure.

Diagnostic Rule

- **Below 20 PSI:** Air in loop or leak in loop field
- **20–30 PSI:** Borderline, may cause performance issues
- **30–60 PSI:** Normal operating range

If pressure is low, the loop must be flushed and re-pressurized by a geothermal technician with a high-volume purge cart.

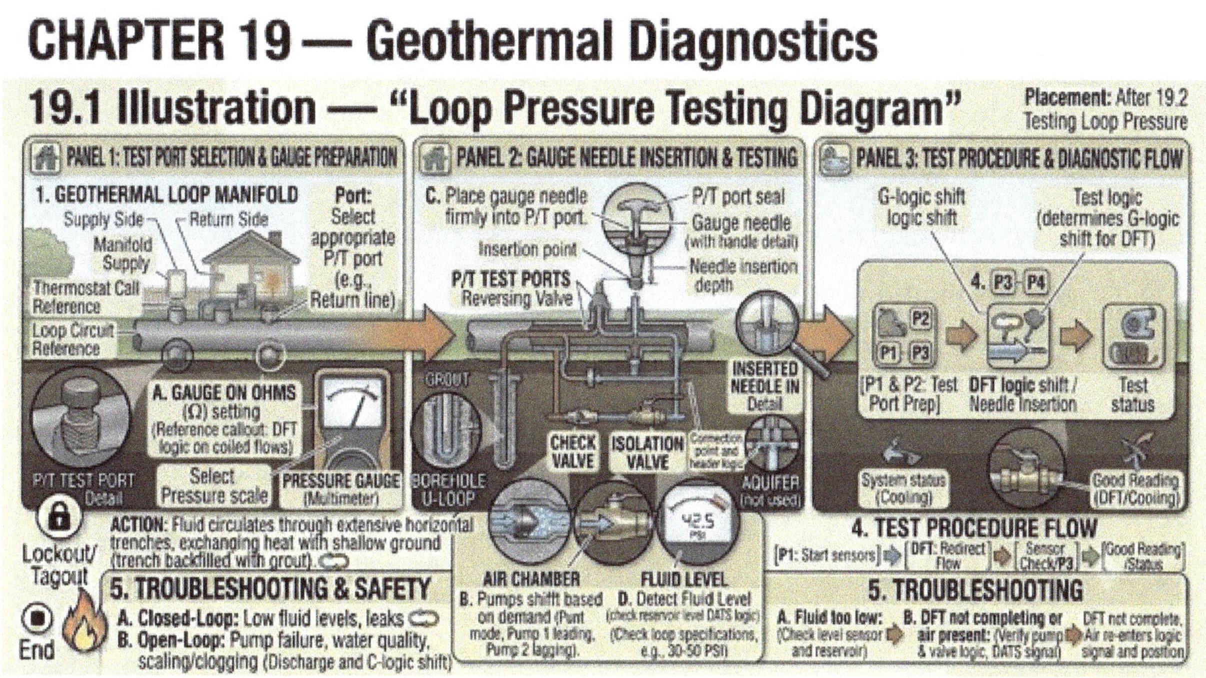

19.3 Testing Circulator Pumps

Circulator pumps are the most common failure point in geothermal systems.

Symptoms of a Failed Pump

- No heating or cooling
- High loop temperature in cooling
- Low loop temperature in heating
- Compressor short cycling
- Flow center making grinding or rattling noises

How to Test a Circulator Pump

1. **Check for power**
   Verify 120V or 240V (depending on model) at the pump terminals.
2. **Check pump amperage**
   Compare to nameplate rating.
   - Zero amps = pump not running
   - Very low amps = seized impeller
   - Very high amps = locked rotor
3. **Check for vibration**
   A running pump will vibrate slightly.
   A silent pump with power applied has failed.
4. **Check temperature difference**
   If the pump is running, the supply and return loop temperatures should differ by 5–15°F depending on mode.

Diagnostic Rule

If the pump has power but does not run, it has failed.

CHAPTER 19 — "Circulator Pump Diagnostic

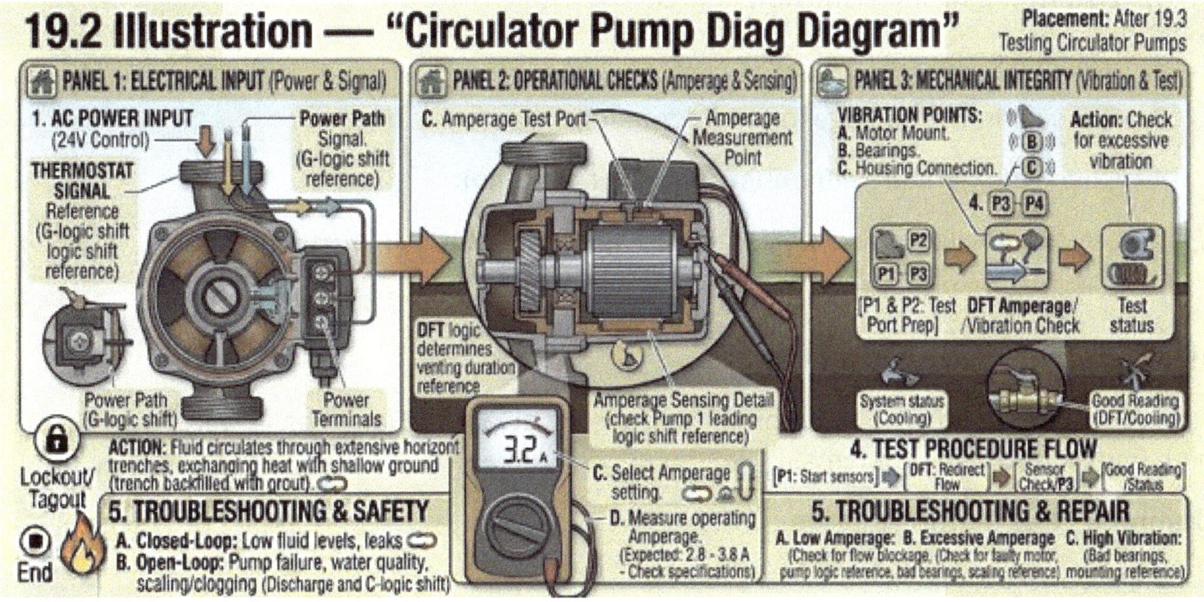

19.4 Diagnosing Air in the Loop

Air is the enemy of geothermal systems. Even a small amount can stop circulation.

Symptoms of Air in the Loop

- Gurgling or bubbling sounds
- Flow center cavitation
- Poor heating or cooling
- Rapid temperature swings
- Low loop pressure

How Air Gets Into the Loop

- Improper installation
- Leaks in underground piping
- Failed O-rings in flow center
- Low antifreeze concentration causing micro-bubbles

How to Diagnose Air in the Loop

1. Listen for gurgling in the flow center.
2. Check loop pressure (usually low).
3. Check pump amperage (often fluctuating).
4. Check temperature difference (erratic or extreme).

If air is present, the loop must be purged using a high-volume flush cart.

19.5 Diagnosing Flow Center Issues

Flow centers contain the circulator pumps, check valves, and isolation valves. Failures here mimic loop field problems.

Common Flow Center Failures

- Stuck check valve
- Failed circulator pump
- Air lock
- Internal leak
- Cavitation

How to Diagnose Flow Center Problems

1. Verify pump operation.
2. Verify loop pressure.
3. Check for temperature rise/fall across the heat pump.
4. Check for vibration or noise from the flow center.
5. Verify check valves are not stuck closed.

Diagnostic Rule

If loop pressure is normal but flow is low, the issue is inside the flow center — not the loop field.

19.6 Diagnosing Open-Loop Water Quality Problems

Open-loop systems use groundwater directly. Water quality is the #1 cause of failure.

Common Water Quality Issues

- Iron buildup
- Calcium scaling
- Sand or sediment
- Biological growth

Symptoms of Water Quality Problems

- Reduced water flow
- High head pressure
- Poor cooling performance
- Heat exchanger fouling
- Compressor overload

How to Diagnose

1. Check water flow rate at discharge.
2. Inspect filters or strainers.
3. Check temperature difference across the heat pump.
4. Look for rust or mineral deposits in piping.

If scaling is present, the heat exchanger may require chemical cleaning by a geothermal technician.

19.7 Using Temperature and Pressure to Diagnose Performance

Geothermal systems provide excellent diagnostic data through simple temperature and pressure readings.

Heating Mode Diagnostic Targets

- Entering loop temperature: 30–50°F
- Leaving loop temperature: 20–40°F
- Temperature rise across air coil: 20–30°F

Cooling Mode Diagnostic Targets

- Entering loop temperature: 60–80°F
- Leaving loop temperature: 70–90°F
- Temperature drop across air coil: 15–20°F

Diagnostic Rule

If loop temperatures are normal but performance is poor, the issue is inside the heat pump. If loop temperatures are abnormal, the issue is in the loop or flow center.

19.8 Geothermal Troubleshooting Matrix

Symptom: No heating or cooling
Likely Causes:

- Failed circulator pump
- Low loop pressure
- Air in loop
- Compressor not running

Symptom: Poor heating performance
Likely Causes:

- Low loop temperature
- Undersized loop field
- Air in loop
- Failed pump

Symptom: Poor cooling performance
Likely Causes:

- High loop temperature
- Scaling (open-loop)
- Flow restriction

Symptom: System locks out
Likely Causes:

- Low flow
- Low pressure
- Thermistor failure
- Compressor overload

Symptom: Gurgling noises
Likely Cause:

- Air in loop

19.9 What You'll Learn Next

With geothermal diagnostics complete, the next section of the book returns to your standard structure:

- Reference tables
- Glossary
- Troubleshooting matrices
- Wiring diagrams
- System behavior charts

These final chapters complete the handbook and give readers a technician-level reference they can use for years.

CHAPTER 20 — Advanced Preventative Maintenance for Heat Pumps & Geothermal Systems

Preventative maintenance is the backbone of HVAC reliability. Whether the system is a standard air-source heat pump or a high-efficiency geothermal unit, the principles are the same: clean airflow, proper electrical performance, and stable operating conditions. This chapter teaches the advanced maintenance procedures that dramatically extend system life, reduce energy consumption, and prevent the most common seasonal failures.

Most homeowners think maintenance is just "changing the filter." Technicians know the truth: real maintenance is a structured, multi-step process that verifies airflow, electrical health, refrigerant behavior (without touching gauges), and system logic.

20.1 Why Maintenance Matters More for Heat Pumps and Geothermal

Heat pumps and geothermal systems run **year-round**, unlike furnaces or air conditioners that only run seasonally. This means:

- More runtime hours
- More wear on motors and capacitors
- More airflow restrictions
- More opportunities for electrical degradation
- More sensitivity to dirt, dust, and coil contamination

A neglected heat pump loses **20–40% efficiency**.
A neglected geothermal system loses **10–25% efficiency**.

Maintenance is not optional — it is essential.

20.2 Seasonal Maintenance Checklist (Air-Source Heat Pumps)

This is the technician-level checklist used in the field.

Indoor Air Handler

- Replace or clean air filter
- Inspect blower wheel for dust buildup
- Test blower motor capacitor
- Inspect ECM module for overheating
- Clean evaporator coil (if accessible)
- Flush condensate drain
- Inspect heat strips for loose connections
- Test sequencers or relays
- Verify thermostat wiring integrity

Outdoor Unit

- Deep-clean condenser coil
- Straighten bent fins
- Test dual-run capacitor (HERM and FAN)
- Test contactor for pitting or carbon buildup
- Inspect reversing valve coil
- Inspect defrost board for corrosion
- Test outdoor thermistor resistance
- Inspect fan motor bearings
- Verify proper refrigerant behavior using temperature split (no gauges)

System-Wide

- Verify 24V transformer output
- Inspect all low-voltage wiring for rub-outs
- Tighten all electrical connections
- Verify thermostat staging (Y1, Y2, W2, O/B)
- Confirm auxiliary heat operation

20.3 Seasonal Maintenance Checklist (Geothermal Systems)

Geothermal systems require additional steps due to the loop field and flow center.

Indoor Heat Pump Unit

- Clean air coil
- Inspect blower wheel
- Test blower motor capacitor or ECM module
- Inspect condensate pan and drain
- Test reversing valve coil
- Test compressor amperage
- Inspect electrical connections

Flow Center

- Verify circulator pump operation
- Check loop pressure
- Inspect for air in loop (listen for gurgling)
- Check for cavitation
- Inspect check valves
- Inspect isolation valves
- Verify temperature difference across loop

Loop Field Diagnostics

- Measure entering loop temperature
- Measure leaving loop temperature
- Compare readings to expected seasonal ranges
- Inspect antifreeze concentration (if accessible)

Open-Loop Systems

- Inspect strainers or filters
- Check for scaling or mineral buildup
- Verify well pump operation
- Inspect discharge line

20.4 Electrical Maintenance: The Hidden Lifespan Killer

Electrical degradation is the #1 cause of heat pump and geothermal failures.

Capacitors

Capacitors weaken gradually. A capacitor that tests “within range” but near the low end can still cause:

- Hard starts
- Overheating
- Compressor failure
- Fan motor failure

Technicians replace weak capacitors proactively.

Contactors

Pitted or carbon-scarred contactors cause:

- Voltage drop
- Overheating
- Compressor stress
- Motor failure

A contactor is a $20 part that protects a $2,000 compressor.

Low-Voltage Wiring

Heat pumps rely heavily on low-voltage signals:

- Y (compressor)
- O/B (reversing valve)
- W2 (aux heat)
- G (blower)

Loose or corroded connections cause intermittent failures that mimic major system problems.

20.5 Airflow Maintenance: The Silent Efficiency Killer

Airflow problems cause:

- Freeze-ups
- Overheating
- High energy bills
- Poor comfort
- Compressor stress

Airflow Maintenance Steps

- Replace filters regularly
- Clean blower wheel
- Clean evaporator coil
- Inspect ductwork for leaks
- Verify return air sizing
- Check static pressure

Static Pressure Targets

- 0.3–0.5 in. w.c. = ideal
- 0.6–0.8 in. w.c. = borderline
- 0.9+ in. w.c. = airflow restriction

High static pressure destroys heat pumps.

20.6 Temperature-Based Refrigerant Diagnostics (No Gauges Needed)

Your book avoids EPA-restricted work, but you can still diagnose refrigerant behavior using temperature readings.

Cooling Mode

- Supply air: 55–60°F
- Return air: 70–78°F
- Temperature drop: 15–20°F

Heating Mode (Air-Source)

- Supply air: 85–100°F
- Return air: 65–75°F
- Temperature rise: 15–30°F

Heating Mode (Geothermal)

- Supply air: 90–110°F
- Return air: 65–75°F
- Temperature rise: 20–30°F

Diagnostic Rule

If airflow is correct but temperature split is abnormal, the issue is refrigerant-side — and requires an EPA-certified technician.

20.7 Preventative Maintenance for Auxiliary Heat

Auxiliary heat is critical in cold weather.

Maintenance Steps

- Inspect heat strips
- Test sequencers
- Test relays
- Verify 240V supply
- Inspect wiring for heat damage
- Test temperature rise with strips energized

Diagnostic Rule

If the home cannot reach setpoint in winter, auxiliary heat is the first suspect.

20.8 Preventative Maintenance for Defrost Systems

Defrost failures are the #1 cause of winter heat pump breakdowns.

Maintenance Steps

- Test outdoor thermistor
- Inspect defrost board
- Verify reversing valve operation
- Inspect outdoor coil for dirt
- Verify outdoor fan operation
- Inspect coil sensor placement

Diagnostic Rule

If the outdoor unit freezes, the defrost system is malfunctioning.

20.9 Annual Maintenance Schedule

Twice Per Year (Spring & Fall)

- Full heat pump inspection
- Full geothermal inspection
- Coil cleaning
- Electrical testing
- Airflow testing

Once Per Year

- Deep condenser cleaning
- Blower wheel cleaning
- Heat strip testing
- Loop field temperature verification

Every 3–5 Years

- Geothermal loop antifreeze testing
- Open-loop heat exchanger cleaning
- Full duct inspection

20.10 What You'll Learn Next

The next chapter returns to your book's standard structure:

- Reference tables
- Glossary
- Wiring diagrams
- Troubleshooting matrices

CHAPTER 21 — Master Troubleshooting Matrix

This chapter is the technician's quick-reference guide. Every HVAC failure — whether heating, cooling, heat pump, or geothermal — can be traced to a specific step in the sequence of operation. The tables below map **symptoms → failed step → likely causes**, allowing you to diagnose systems without guessing.

21.1 Furnace Troubleshooting Matrix (Gas Heat)

Symptom: Thermostat clicks, but furnace is completely silent
Sequence Step Failed: Step 1 or 2
Likely Causes:

- No 24V between R and C
- Blown 3A/5A low-voltage fuse
- Open high-limit or rollout switch
- Failed 24V transformer
- Broken thermostat wire

Symptom: Inducer runs, but igniter never glows
Sequence Step Failed: Step 4
Likely Causes:

- Clogged pressure switch port
- Cracked pressure switch hose
- Blocked flue or intake
- Failed pressure switch

Symptom: Igniter glows, but no flame
Sequence Step Failed: Step 6
Likely Causes:

- Gas valve not receiving 24V
- Manual gas shutoff closed
- Failed gas valve
- Low gas pressure

Symptom: Burners light, then shut off after 3–5 seconds
Sequence Step Failed: Step 8
Likely Causes:

- Dirty flame sensor
- Loose ground
- Failed control board flame circuit

Symptom: Furnace heats, then shuts down early
Sequence Step Failed: Step 9
Likely Causes:

- Clogged filter
- Dirty blower wheel
- Weak blower capacitor
- Blocked return air
- Overheating heat exchanger

21.2 Air Conditioner Troubleshooting Matrix (Cooling Mode)

Symptom: AC won't turn on at all
Sequence Step Failed: Step 1
Likely Causes:

- No 24V between R and Y
- Bad thermostat
- Broken thermostat wire
- Blown low-voltage fuse
- Failed transformer

Symptom: Indoor blower runs, outdoor unit dead
Sequence Step Failed: Step 3
Likely Causes:

- No 24V at contactor
- Broken low-voltage wire outside
- Failed contactor coil
- Safety switch open

Symptom: Contactor pulls in, but compressor and fan don't run
Sequence Step Failed: Step 4
Likely Causes:

- Failed dual-run capacitor
- Failed compressor
- Failed condenser fan motor
- Burned contactor points
- Low line voltage

Symptom: Compressor hums but won't start
Likely Causes:

- Weak or failed capacitor
- Locked compressor
- Low voltage
- Failed start winding

Symptom: AC runs but doesn't cool well
Likely Causes:

- Dirty condenser coil
- Dirty evaporator coil
- Low airflow
- Low refrigerant (diagnosis only)

21.3 Heat Pump Troubleshooting Matrix (Heating Mode)

Symptom: Heat pump blows cold air
Sequence Step Failed: Reversing valve or compressor operation
Likely Causes:

- Reversing valve stuck
- No O/B signal
- Outdoor unit not running
- Defrost board stuck in cooling
- Low refrigerant (diagnosis only)

Symptom: Outdoor unit frozen solid
Likely Causes:

- Defrost board failure
- Outdoor thermistor failure
- Outdoor fan not running
- Low airflow across outdoor coil

Symptom: Heat pump runs constantly
Likely Causes:

- Auxiliary heat not engaging
- Undersized system
- Low outdoor temperature
- Dirty indoor coil
- Weak blower motor

Symptom: Loud "whoosh" sound
Likely Cause:

- Reversing valve shifting (normal)

Symptom: Heat pump heats in cooling mode
Likely Causes:

- Reversing valve coil failure
- Incorrect O/B thermostat setting
- Defrost board mis-energizing valve

21.4 Auxiliary Heat Troubleshooting Matrix

Symptom: Home cannot reach setpoint in winter
Likely Causes:

- Heat strips not energizing
- Failed sequencer
- Failed relay
- Broken W2 wire
- Thermostat not sending W2

Symptom: Burning smell when strips energize
Likely Causes:

- Dust on heat strips (normal)
- Loose electrical connection (dangerous)
- Overheating element

21.5 Defrost System Troubleshooting Matrix

Symptom: Outdoor unit never defrosts
Likely Causes:

- Failed defrost board
- Failed outdoor thermistor
- Reversing valve not shifting
- Outdoor fan stuck running

Symptom: Unit defrosts too often
Likely Causes:

- Bad thermistor
- Dirty outdoor coil
- Low airflow
- Undersized outdoor unit

21.6 Geothermal Troubleshooting Matrix

Symptom: No heating or cooling
Likely Causes:

- Failed circulator pump
- Low loop pressure
- Air in loop
- Compressor not running

Symptom: Poor heating performance
Likely Causes:

- Low loop temperature
- Undersized loop field
- Air in loop
- Failed pump

Symptom: Poor cooling performance
Likely Causes:

- High loop temperature
- Scaling (open-loop)
- Flow restriction

Symptom: Gurgling noises
Likely Cause:

- Air in loop

Symptom: System locks out
Likely Causes:

- Low flow
- Low pressure
- Thermistor failure
- Compressor overload

21.7 Electrical Failure Matrix (All Systems)

Symptom: System dead, no response
Likely Causes:

- Tripped breaker
- Failed transformer
- Blown low-voltage fuse
- Open safety switch

Symptom: Intermittent operation
Likely Causes:

- Loose low-voltage wire
- Corroded spade connector
- Weak capacitor
- Failing contactor

Symptom: Burning smell from equipment
Likely Causes:

- Overheating motor
- Failing capacitor
- Loose high-voltage connection
- Shorted winding

21.8 Airflow Failure Matrix (All Systems)

Symptom: Weak airflow
Likely Causes:

- Dirty filter
- Dirty blower wheel
- Undersized ductwork
- Blocked return
- Weak blower capacitor

Symptom: Evaporator coil freezing
Likely Causes:

- Low airflow
- Dirty coil
- Blower not running
- Low refrigerant (diagnosis only)

Symptom: High energy bills
Likely Causes:

- High static pressure
- Dirty coils
- Long runtimes due to airflow restriction

21.9 Thermostat Troubleshooting Matrix

Symptom: System won't turn on
Likely Causes:

- Dead batteries
- Broken R or C wire
- Miswired thermostat
- Blown low-voltage fuse

Symptom: Heat pump stuck in wrong mode
Likely Causes:

- Incorrect O/B setting
- Reversing valve not energized
- Thermostat misconfigured

Symptom: System short cycles
Likely Causes:

- Incorrect cycle rate setting
- Loose thermostat wire
- Poor thermostat placement

CHAPTER 22 — Wiring Diagrams & Control Logic Reference

Understanding HVAC wiring is the key to diagnosing electrical failures quickly and safely. This chapter provides clear, technician-level wiring references for furnaces, air conditioners, heat pumps, and geothermal systems. These diagrams are simplified for clarity and designed to match the diagnostic methods taught throughout this handbook.

22.1 Thermostat Terminal Reference (Universal Guide)

Every modern thermostat uses the same core terminals. Not all systems use every terminal, but all terminals follow the same logic.

R — 24V power (hot)
C — 24V common
Y — Compressor (stage 1)
Y2 — Compressor (stage 2)
G — Indoor blower
W — Heat (furnace)
W2/AUX — Auxiliary heat (heat pump)
O/B — Reversing valve (heat pump)
L — System monitor / fault indicator
E — Emergency heat
S1/S2 — Outdoor temperature sensor
RH/RC — Separate heating/cooling transformers (jumpered in most homes)

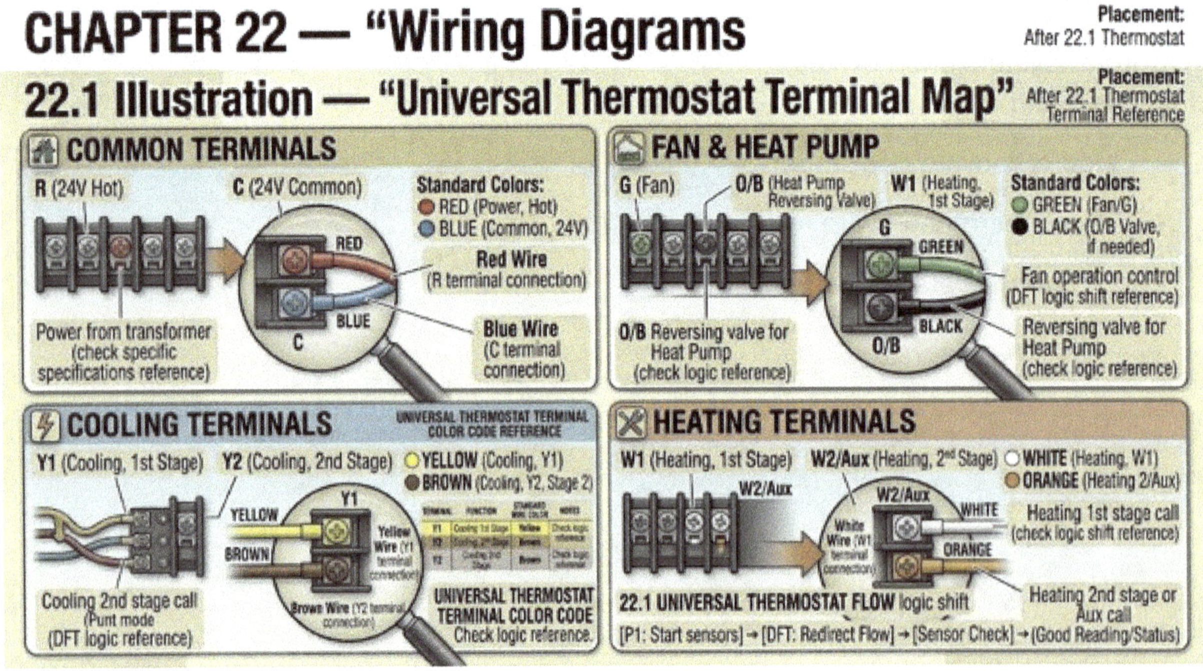

22.2 Standard Furnace Wiring Diagram (Simplified)

Thermostat → Furnace Control Board

- R → R (24V power)
- C → C (common)
- W → W (heat call)
- G → G (blower)
- Y → Y (cooling signal passed to outdoor unit)

Furnace Control Board → Components

- W → Gas valve
- G → Blower relay
- Y → Outdoor contactor
- Limit switches → Safety loop
- Transformer → R and C supply

Diagnostic Rule:
If the furnace is dead, check R → C for 24V first.

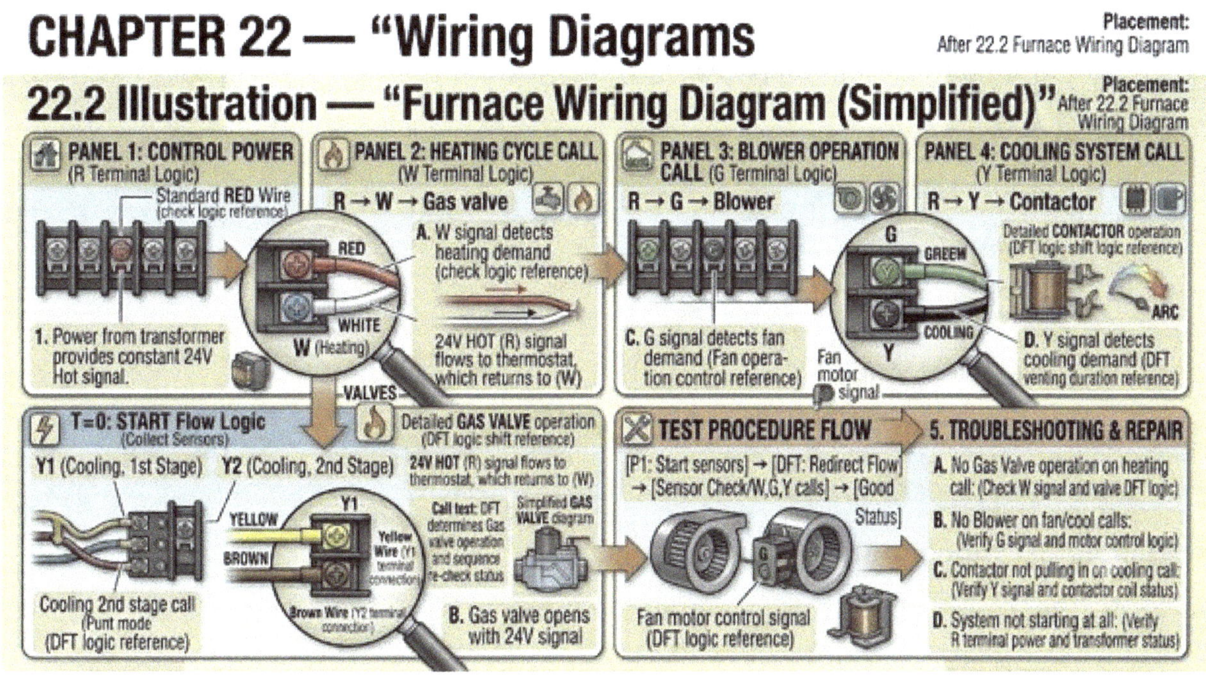

22.3 Standard Air Conditioner Wiring Diagram

Thermostat → Furnace/Air Handler

- Y → Compressor contactor
- G → Indoor blower
- R/C → 24V power

Furnace/Air Handler → Outdoor Unit

- Y → Contactor coil
- C → Contactor coil

Outdoor Unit Internal Wiring

- Contactor → Compressor
- Contactor → Fan motor
- Capacitor → Compressor (HERM)
- Capacitor → Fan (FAN)

Diagnostic Rule:
If the contactor is pulled in but nothing runs, suspect the capacitor or high-voltage circuit.

22.4 Heat Pump Wiring Diagram (O/B Logic)

Thermostat → Air Handler

- Y → Compressor
- G → Blower
- O/B → Reversing valve
- AUX/W2 → Heat strips
- E → Emergency heat
- R/C → 24V power

Air Handler → Outdoor Unit

- Y → Contactor
- C → Contactor
- O/B → Reversing valve coil

Diagnostic Rule:
If the system heats when it should cool, check the O/B setting first.

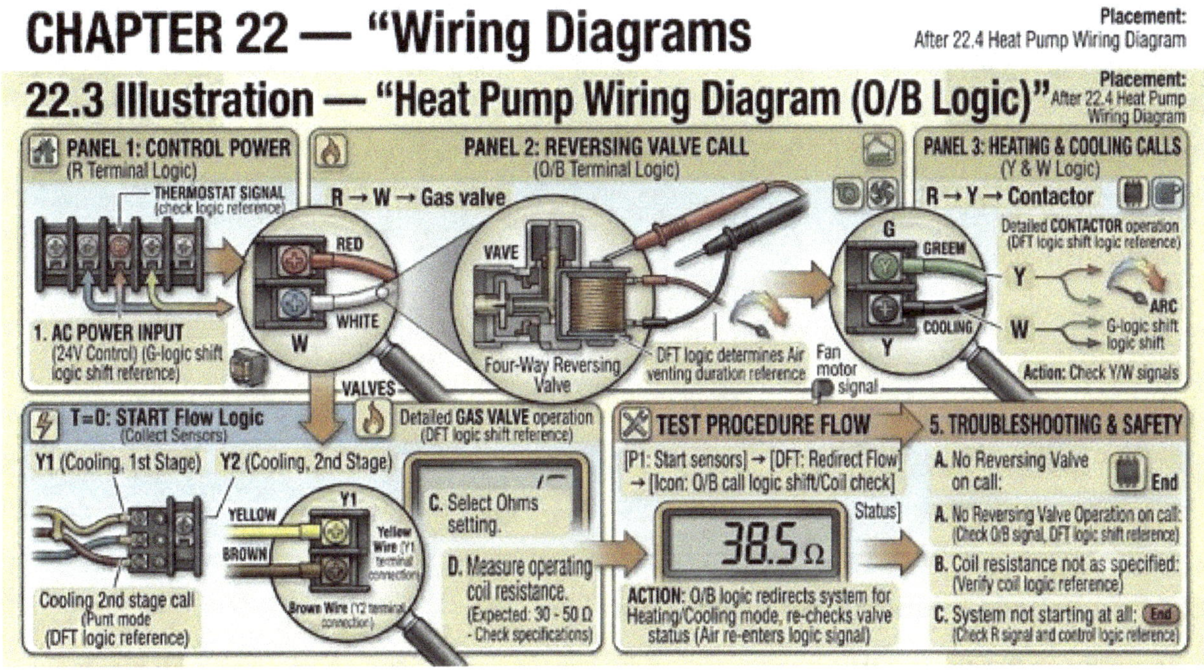

22.5 Geothermal Heat Pump Wiring Diagram

Thermostat → Geothermal Unit

- Y → Compressor
- G → Blower
- O/B → Reversing valve
- AUX/W2 → Electric backup heat
- R/C → 24V power

Geothermal Unit → Flow Center

- Pump relay → Circulator pump(s)
- Safety switches → Lockout circuit

Diagnostic Rule:
If the geothermal unit runs but produces no heating or cooling, check circulator pump operation.

CHAPTER 22 — "Wiring Diagrams

Placement:
After 22.5 Geothermal Wiring Diagram

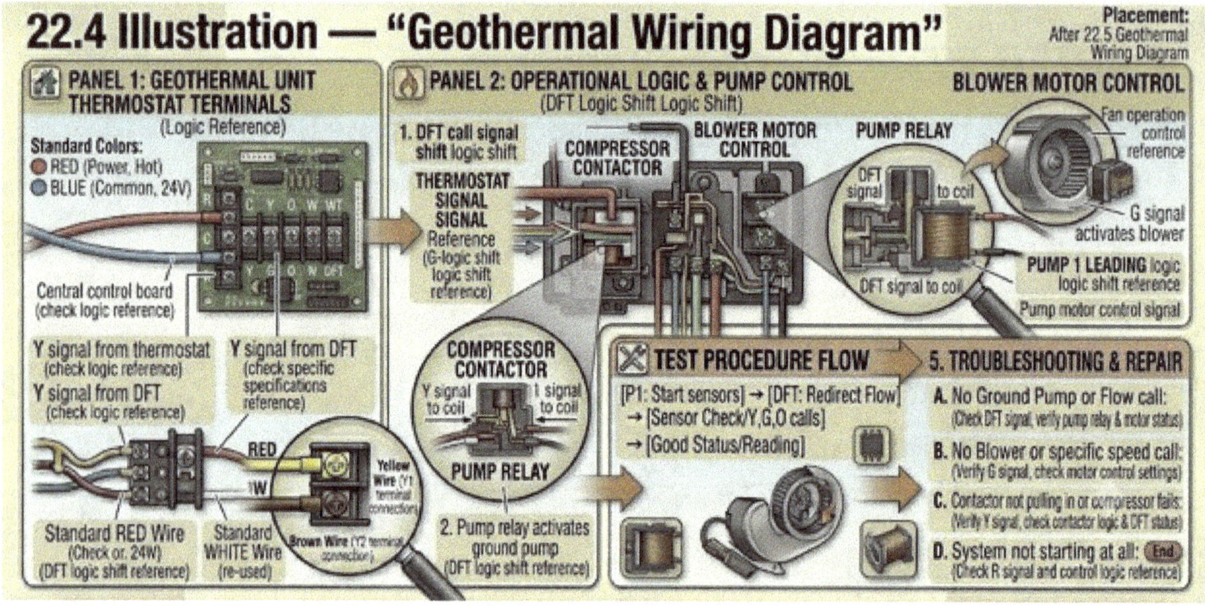

22.6 Low-Voltage Troubleshooting Flowchart

1. **Check R → C for 24V**
   - No voltage: transformer or fuse failure
2. **Check thermostat output (R → Y, R → W, R → G)**
   - No output: thermostat or wiring failure
3. **Check control board output**
   - No output: board failure
4. **Check component voltage**
   - Voltage present but no operation: component failure
5. **Check safety switches**
   - Open switch stops sequence

22.7 High-Voltage Path Diagrams

Furnace

Breaker → Switch → Control board → Blower motor / Igniter / Gas valve

Air Conditioner

Breaker → Disconnect → Contactor → Compressor / Fan motor

Heat Pump

Breaker → Disconnect → Contactor → Compressor / Fan motor / Reversing valve

Geothermal

Breaker → Unit → Compressor / Blower / Pump relay

www.ingramcontent.com/pod-product-compliance
Lightning Source LLC
LaVergne TN
LVHW060036210926
848481LV00013B/3877
* 9 7 9 8 9 9 6 6 4 6 9 3 7 *